AF614740

Time Frequency Analysis of Some Generalized Fourier Transforms

Edited by Mohammad Younus Bhat

Published in London, United Kingdom

Time Frequency Analysis of Some Generalized Fourier Transforms
http://dx.doi.org/10.5772/intechopen.104023
Edited by Mohammad Younus Bhat

Contributors
Mohammad Younus Bhat, Aamir Hamid Dar, Altaf Ahmad Bhat, Deepak Kumar Jain, Miodrag D. Kušljević, Didar Urynbassarova, Altyn Urynbassarova, Showkat Ahmad Dar, M. Kamarujjama

First published in London, United Kingdom, 2023 by IntechOpen
IntechOpen is the global imprint of INTECHOPEN LIMITED, registered in England and Wales, registration number: 11086078, 5 Princes Gate Court, London, SW7 2QJ, United Kingdom

British Library Cataloguing-in-Publication Data
A catalogue record for this book is available from the British Library

Additional hard and PDF copies can be obtained from orders@intechopen.com

Time Frequency Analysis of Some Generalized Fourier Transforms
Edited by Mohammad Younus Bhat
p. cm.
Print ISBN 978-1-83768-459-5
Online ISBN 978-1-83768-460-1
eBook (PDF) ISBN 978-1-83768-461-8

For EU product safety concerns:
IN TECH d.o.o., Prolaz Marije Krucifikse Kozulić 3, 51000 Rijeka, Croatia,
info@intechopen.com or visit our website at intechopen.com.

Meet the editor

Dr. Mohammad Younus Bhat is an assistant professor in the Department of Mathematical Sciences, Islamic University of Science and Technology, Kashmir. He obtained a master's degree in Mathematics from the University of Kashmir and a doctorate degree in Mathematics from the Central University of Jammu. He has more than sixty research papers and five book chapters to his credit. His prime research areas are signal and image processing, harmonic analysis, wavelet analysis, numerical analysis, and differential equations.

Contents

Preface

Joseph Fourier (1770–1830) first introduced the remarkable idea of expansion of a function in terms of trigonometric series without giving any attention to rigorous mathematical analysis. The integral formulas for the coefficients of the Fourier expansion were already known to Leonardo Euler (1707–1783) and others. In fact, Fourier developed his new idea for finding the solution of heat (or Fourier) equation in terms of Fourier series so that the Fourier series can be used as a practical tool for determining the Fourier series solution of partial differential equations under prescribed boundary conditions.

The Fourier transform originated from the Fourier integral theorem that was stated in the Fourier treatise titled *La Théore Analytique de la Chaleur,* and its deep significance has subsequently been recognized by mathematicians and physicists. It is generally believed that the theory of Fourier series and Fourier transforms is one of the most remarkable discoveries in mathematical sciences and it has widespread applications in mathematics, physics, and engineering. Both the Fourier series and Fourier transforms are related in many important ways. Many applications, including the analysis of stationary signals and real-time signal processing, make effective use of the Fourier transform in time and frequency domains.

In time-frequency analysis, the Fourier transform is one of the oldest tools to dominate signal processing. However, due to its drawbacks in the analysis of non-stationary signals, different alternative transforms have gained much popularity in recent years, including windowed Fourier transform, fractional Fourier transform, linear canonical transform, quadratic-phase Fourier transform, and so on. These transforms are known as generalizations of the classic Fourier transform.

The main reason for writing this book is to stimulate interactions among mathematicians, computer scientists, engineers, and economists, as well as biological and physical scientists. The text is suitable for advanced graduate students but is primarily intended for post-graduate students and researchers in wavelets and their applications.

The book begins with an elementary chapter that introduces general Fourier transforms like windowed Fourier transform, fractional Fourier transform, linear canonical transform, and quadratic-phase Fourier transform.

Hybrid transforms are constructed by associating the Wigner-Ville distribution (WVD) with widely known signal processing tools, such as fractional Fourier transform, linear canonical transform, offset linear canonical transform (OLCT), and their quaternion-valued versions. Chapter 2 summarizes research on hybrid transforms by reviewing a computationally efficient type of WVD-OLCT, which has simplicity in marginal properties compared to classic WVD-OLCT and WVD.

Quadratic-phase Fourier transform (QPFT) as a general integral transform has been generalized into Wigner distribution (WD) and ambiguity function (AF) to show a more powerful ability for non-stationary signal processing. Chapter 3 proposes a new version of AF associated with QPFT referred to as scaled AF. This new version of AF is defined based on the QPFT and the fractional instantaneous autocorrelation.

Chapter 4 presents analytical expressions of infinite Fourier sine and cosine transform-based Ramanujan integrals in an infinite series of hypergeometric functions using the hypergeometric technique. Moreover, as applications of Ramanujan's integrals, some closed form of infinite summation formulae involving hypergeometric functions are derived.

Chapter 5 is devoted to the recursive algorithms for harmonic analysis, one of which is the resonator-based algorithm. The approach of the parallel cascades of multiple-resonators (MRs) with the common feedback is generalized as the cascaded-resonator (CR)-based structure for recursive harmonic analysis.

Dr. Mohammad Younus Bhat
Department of Mathematical Sciences,
Islamic University of Science and Technology,
Kashmir, India

Section 1

Transforms

Chapter 1

Introductory Chapter: The Generalizations of the Fourier Transform

Mohammad Younus Bhat

1. Introduction

In the world of physical science, important physical quantities such as sound, pressure, electric current, voltage, and electromagnetic fields vary with time t. Such quantities are labeled as signals/waveforms. Exemplified by signals with examples such as oral signals, optical signals, acoustic signals, biomedical signals, radar, and sonar. Indeed, signals are very common in the real world. Time-frequency analysis is a vital aid in signal analysis, which is concerned with how the frequency of a function (or signal) behaves in time, and it has evolved into a widely recognized applied discipline of signal processing. The signals can be classified under various categories. It could be done in terms of continuity (continuous v/s discrete), periodicity(periodic v/s aperiodic), stationarity(stationary v/s non-stationary), and so on. Most of the signals in nature are non-stationary (i.e., whose spectral components change with time) and apt presentation of such non-stationary signals need frequency analysis, which is local in time, resulting in the time-frequency analysis of signals. Although time frequency analysis of signals had its origin almost 70 years ago, there has been major development of the time-frequency distribution approach in the last three decades. The basic idea of these methods is to develop a joint function of time and frequency, known as a time-frequency distribution, that can describe the energy density of a signal simultaneously in both time and frequency domains. In signal processing, time-frequency analysis comprises those techniques that study signal in both the time and frequency domains simultaneously, using various time-frequency representations/tools known as integral transformations. An integral transform maps a function/signal from one function space into another function space *via* integration, where some of the properties of the original function might be more easily characterized and manipulated than in the original function space. The integral transforms are essentially considered from the functional analysis viewpoint and as a useful technique of mathematical physics.

The classical Fourier transform (FT) is an integral transform introduced by Joseph Fourier in 1807 [1], is one of the most valuable and widely-used integral transforms that converts a signal from time versus amplitude to frequency versus amplitude. Thus FT can be considered as the time-frequency representation tool in signal processing and analysis. A fundamental limitation of the Fourier transform is that the all properties of a signal are global in scope. Information about local features of the signal, such as changes in frequency, becomes a global property of the signal in the frequency domain. In order to circumvent these drawbacks of FT, authors in Ref. [2] introduced the generalizations of FT that includes short-time Fourier transform (STFT) by performing the FT on a block-by-block basis rather than to process the entire signal at once. In spite of the fact that STFT did much to ameliorate the limitations of FT, still in some cases the STFT cannot track the signal dynamics properly for a signal with both very high frequencies of short duration and very low frequencies of long duration. To overcome these drawbacks of FT and STFT different novel generalizations of the classical Fourier transform came into existence *viz.*: the fractional Fourier transform (FRFT), the Fresnal transform, the linear canonical transform (LCT), the quadratic-phase Fourier transform (QPFT), and so on. As a generalization of classical Fourier transform, the FRFT, the LCT, the QPFT gained its ground intermittently and profoundly influenced several branches of science and engineering including signal and image processing, quantum mechanics, neural networks, differential equations, optics, pattern recognition, radar, sonar, and communication systems.

2. Fourier transform and its generalizations

2.1 Fourier transform

Joseph Fourier [1] in 1822 published first work about Fourier transform, which is an integral transform that converts a mathematical function from the time domain to the frequency domain. Fourier transform measures the frequency component of a given function. The Fourier transform has evolved into a widely recognized discipline of harmonic analysis and has been successfully applied in diverse scientific and engineering pursuits [3–6].

Let us begin with definition of the classical Fourier transform.

Definition 1. *The FT of any signal* $x(t) \in L^2(\mathbb{R})$ *is defined and denoted as*

$$\mathcal{F}[x(t)](\xi) = \hat{x}(\xi) = \frac{1}{\sqrt{2\pi}} \int_{\mathbb{R}} e^{-i\xi t} x(t)dt, \tag{1}$$

and corresponding inversion formula is given by

$$\mathcal{F}^{-1}(\mathcal{F}[x(t)](\xi))(t) = \frac{1}{\sqrt{2\pi}} \int_{\mathbb{R}} e^{i\xi t} \mathcal{F}[x(t)](\xi)d\xi. \tag{2}$$

Example 1. *Consider a function* $x(t) = \begin{cases} e^{-\alpha t} & \text{for } \; t \geq 0, \alpha > 0, \\ 0 & \text{otherwise}; \end{cases}$ *, then the Fourier transform of* $x(t)$ *is obtained as*

$$\begin{aligned} \mathcal{F}[x(t)](\xi) &= \frac{1}{\sqrt{2\pi}} \int_0^\infty e^{-i\xi t} e^{-\alpha t} dt \\ &= \frac{1}{\sqrt{2\pi}} \int_0^\infty (\cos \xi t - i \sin \xi t) e^{-\alpha t} dt \\ &= \frac{1}{\sqrt{2\pi}} \left\{ \int_0^\infty \cos \xi t e^{-\alpha t} dt - i \int_0^\infty \sin \xi t e^{-\alpha t} dt \right\} \\ &= \frac{1}{\sqrt{2\pi}} \left\{ \frac{\alpha}{\alpha^2 + \xi^2} - \frac{i\xi}{\alpha^2 + \xi^2} \right\}. \end{aligned}$$

Example 2. *Consider the function*

$$x(t) = \begin{cases} \sin 3t & \text{for} \quad -\pi \le t \le \pi, \\ 0 & \text{otherwise}. \end{cases}$$

Then the Fourier transform of $x(t)$ is obtained as

$$\begin{aligned} \mathscr{F}[x(t)](\xi) &= \frac{1}{\sqrt{2\pi}} \int_{\mathbb{R}} (\cos \xi t - i \sin \xi t) \sin 3t dt \\ &= \frac{-i}{\sqrt{2\pi}} \int_{-\pi}^{\pi} \sin \xi t \sin 3t dt \\ &= \frac{i 3\sqrt{2} \sin \xi\pi}{\sqrt{\pi}(\xi^2 - 9)}. \end{aligned}$$

Next, we shall study some properties of FT.

Theorem 1 (Translation). *The Fourier transform of any function $x(t-k)$ is given by*

$$\mathscr{F}[x(t-k)](\xi) = e^{-i\xi k} \mathscr{F}[x(t)](\xi). \tag{3}$$

Proof. From Definition 1, we have

$$\begin{aligned} \mathscr{F}[x(t-k)](\xi) &= \frac{1}{\sqrt{2\pi}} \int_{\mathbb{R}} e^{-i\xi t} x(t-k) dt \\ &= \frac{1}{\sqrt{2\pi}} \int_{\mathbb{R}} e^{-i\xi(u+k)} x(u) du \\ &= \frac{1}{\sqrt{2\pi}} \int_{\mathbb{R}} e^{-i\xi k} e^{-i\xi u)} x(u) du \\ &= \frac{1}{\sqrt{2\pi}} e^{-i\xi k} \int_{\mathbb{R}} e^{-i\xi u)} x(u) du \\ &= e^{-i\xi k} \mathscr{F}[x(t)](\xi). \end{aligned}$$

This completes the proof. □

Theorem 2 (Modulation). *The Fourier transform of any function $e^{i\xi_0 t} x(t)$ is given by*

$$\mathscr{F}\left[e^{i\xi_0 t} x(t)\right](\xi) = \mathscr{F}[x(t)](\xi - \xi_0). \tag{4}$$

Proof. From Definition 1, we have

$$\begin{aligned} \mathscr{F}\left[e^{i\xi_0 t} x(t)\right](\xi) &= \frac{1}{\sqrt{2\pi}} \int_{\mathbb{R}} e^{-i\xi t} e^{i\xi_0 t} x(t) dt \\ &= \frac{1}{\sqrt{2\pi}} \int_{\mathbb{R}} e^{-i(\xi - \xi_0)t} x(t) dt \\ &= \mathscr{F}[x(t)](\xi - \xi_0). \end{aligned}$$

This completes the proof. □

Theorem 3 (Orthogonality relation). *The Fourier transform of the functions $x(t)$ and $y(t)$ in $L^2(\mathbb{R})$ satisfies the following orthogonality relation*

$$\langle \mathscr{F}[x(t)], \mathscr{F}[y(u)] \rangle = \langle x(t), y(u) \rangle. \tag{5}$$

Proof. We have

$$\begin{aligned}
\langle \mathscr{F}[x(t)], \mathscr{F}[y(u)] \rangle &= \int_{\mathbb{R}} \mathscr{F}[x(t)](\xi)\overline{\mathscr{F}[y(u)](\xi)}d\xi \\
&= \int_{\mathbb{R}} \mathscr{F}[x(t)](\xi)\overline{\left(\frac{1}{\sqrt{2\pi}}\int_{\mathbb{R}} e^{-i\xi u}y(u)du\right)}d\xi \\
&= \int_{\mathbb{R}} \left(\frac{1}{\sqrt{2\pi}}\int_{\mathbb{R}} e^{-i\xi t}x(t)dt\right)\left(\frac{1}{\sqrt{2\pi}}\int_{\mathbb{R}} e^{i\xi u}\overline{y(u)}du\right)d\xi \\
&= \int_{\mathbb{R}^2} x(t)\overline{y(u)}\left(\frac{1}{2\pi}\int_{\mathbb{R}} e^{i\xi(u-t)}d\xi\right)dtdu \\
&= \int_{\mathbb{R}}\int_{\mathbb{R}} x(t)\overline{y(u)}\delta(u-t)dtdu \\
&= \int_{\mathbb{R}} x(t)\overline{y(u)}dt \\
&= \langle x(t), y(u) \rangle.
\end{aligned}$$

This completes the proof. □

Note: If we take $x(t) = y(t)$, the orthogonality relation yields Plancherel's Theorem for the Fourier transforms that states the energy of a signal ln the time domain, is the same as the energy in the frequency domain given as

$$\|\mathscr{F}(x(t))\| = \|x(t)\|. \tag{6}$$

Next, we show that the inverse Fourier operator is the adjoint of the Fourier operator.

Theorem 4. *Let $x(t)$ and $y(t)$ in $L^2(\mathbb{R})$, then*

$$\langle \mathscr{F}[x(t)](\xi), y(\xi) \rangle = \left\langle x(t), \mathscr{F}^{-1}[y](t) \right\rangle. \tag{7}$$

Proof. We have

$$\begin{aligned}
\langle \mathscr{F}[x(t)], y(t) \rangle &= \int_{\mathbb{R}} \mathscr{F}[x(t)](\xi)\overline{y(\xi)}d\xi \\
&= \int_{\mathbb{R}} \left(\frac{1}{\sqrt{2\pi}}\int_{\mathbb{R}} e^{-i\xi t}x(t)dt\right)\overline{y(\xi)}d\xi \\
&= \int_{\mathbb{R}} x(t)\left(\frac{1}{\sqrt{2\pi}}\int_{\mathbb{R}} e^{-i\xi t}\overline{y(\xi)}d\xi\right)dt \\
&= \int_{\mathbb{R}} x(t)\overline{\left(\frac{1}{\sqrt{2\pi}}\int_{\mathbb{R}} e^{i\xi t}y(\xi)d\xi\right)}dt \\
&= \int_{\mathbb{R}} x(t)\overline{\mathscr{F}^{-1}[y](t)}dt \\
&= \left\langle x(t), \mathscr{F}^{-1}[y](t) \right\rangle.
\end{aligned}$$

This completes the proof. □

Theorem 5. *Let $x(t)$ and $y(t)$ in $L^2(\mathbb{R})$, then*

$$\mathcal{F}[(x*y)](\xi) = \sqrt{2\pi}\,\mathcal{F}[x(t)](\xi)\mathcal{F}[y(t)](\xi), \tag{8}$$

where $x*y$ denotes the convolution of the functions $x(t)$ and $y(t)$ and is given by

$$(x*y)(t) = \int_{\mathbb{R}} x(t)y(u-t)dt.$$

Proof. By applying definition of Fourier transform to the convolution of the functions $x(t)$ and $y(t)$, we obtain

$$\begin{aligned}
\mathcal{F}[(x*y)](\xi) &= \frac{1}{\sqrt{2\pi}}\int_{\mathbb{R}}(x*y)(u)e^{-i\xi u}du \\
&= \frac{1}{\sqrt{2\pi}}\int_{\mathbb{R}}\left(\int_{\mathbb{R}} x(t)y(u-t)dt\right)e^{-i\xi u}du \\
&= \frac{1}{\sqrt{2\pi}}\int_{\mathbb{R}}\int_{\mathbb{R}} x(t)y(v)e^{-i\xi(t+v)}dvdt \\
&= \frac{1}{\sqrt{2\pi}}\int_{\mathbb{R}}\int_{\mathbb{R}} e^{-i\xi t}x(t)y(v)e^{-i\xi v}dvdt \\
&= \sqrt{2\pi}\left\{\frac{1}{\sqrt{2\pi}}\int_{\mathbb{R}} e^{-i\xi t}x(t)dt\right\}\left\{\frac{1}{\sqrt{2\pi}}\int_{\mathbb{R}} e^{-i\xi v}y(v)dv\right\} \\
&= \sqrt{2\pi}\mathcal{F}[x(t)](\xi)\mathcal{F}[y(t)](\xi).
\end{aligned}$$

This completes the proof. □

2.2 Windowed Fourier transform

Definition 2. *Let Ψ be a given window function in $L^2(\mathbb{R})$, then the window Fourier transform (WFT) of any function $x(t) \in L^2(\mathbb{R})$ is defined and denoted as*

$$\mathcal{V}_\Psi[x(t)](b,\xi) = \frac{1}{\sqrt{2\pi}}\int_{\mathbb{R}} e^{-i\xi t}x(t)\overline{\Psi(t-b)}dt, \quad b,\xi \in \mathbb{R}. \tag{9}$$

Further, the WFT (9) can be rewritten as

$$\mathcal{V}_\Psi[x(t)](b,\xi) = \mathcal{F}\left[x(t)\overline{\Psi(t-b)}\right]. \tag{10}$$

Applying inverse FT (2), (10) yields

$$\begin{aligned}
x(t)\overline{\Psi(t-b)} &= \mathcal{F}^{-1}[\mathcal{V}_\Psi[x(t)](b,\xi)] \\
&= \frac{1}{\sqrt{2\pi}}\int_{\mathbb{R}} e^{i\xi t}\mathcal{V}_\Psi[x(t)](b,\xi)d\xi
\end{aligned} \tag{11}$$

Multiplying (11) both sides by $\Psi(t-b)$ and then integrating with respect to db, we get

$$x(t)\|\Psi\|^2=\frac{1}{\sqrt{2\pi}}\int_{\mathbb{R}}\int_{\mathbb{R}}e^{i\xi t}\mathcal{V}_\Psi[x(t)](b,\xi)\Psi(t-b)d\xi db.$$

Equivalently, we have

$$x(t)=\frac{1}{\sqrt{2\pi}\|\Psi\|^2}\int_{\mathbb{R}}\int_{\mathbb{R}}e^{i\xi t}\mathcal{V}_\Psi[x(t)](b,\xi)\Psi(t-b)d\xi db. \tag{12}$$

Eq. (12) gives the inversion formula corresponding to WFT (9).

Theorem 6 (Orthogonality relation). *For any two functions $x(t), y(t)$ in $L^2(\mathbb{R})$, we have following relation*

$$\langle\mathcal{V}_\Psi[x(t)](b,\xi),\mathcal{V}_\Psi[y(t)](b,\xi)\rangle=\|\Psi\|^2\langle x(t),y(t)\rangle. \tag{13}$$

Proof. By Definition (2), we have

$$\begin{aligned}
&\langle\mathcal{V}_\Psi[x(t)](b,\xi),\mathcal{V}_\Psi[y(t)](b,\xi)\rangle\\
&=\int_{\mathbb{R}}\int_{\mathbb{R}}\mathcal{V}_\Psi[x(t)](b,\xi)\overline{\mathcal{V}_\Psi[y(t)](b,\xi)}d\xi db\\
&=\int_{\mathbb{R}}\int_{\mathbb{R}}\mathcal{V}_\Psi[x(t)](b,\xi)\overline{\left(\frac{1}{\sqrt{2\pi}}\int_{\mathbb{R}}e^{-i\xi t}y(t)\overline{\Psi(t-b)}dt\right)}d\xi db\\
&=\int_{\mathbb{R}}\int_{\mathbb{R}}\left(\frac{1}{\sqrt{2\pi}}\int_{\mathbb{R}}e^{i\xi t}\mathcal{V}_\Psi[x(t)](b,\xi)d\xi\right)\overline{y(t)}\Psi(t-b)dtdb.
\end{aligned} \tag{14}$$

By virtue of Eq. (11), (14) yields

$$\begin{aligned}
&\langle\mathcal{V}_\Psi[x(t)](b,\xi),\mathcal{V}_\Psi[y(t)](b,\xi)\rangle\\
&=\int_{\mathbb{R}}\int_{\mathbb{R}}x(t)\overline{\Psi(t-b)}\Psi(t-b)\overline{y(t)}dtdb\\
&=\int_{\mathbb{R}}x(t)\overline{y(t)}dt\int_{\mathbb{R}}\overline{\Psi(t-b)}\Psi(t-b)db\\
&=\|\Psi\|^2\langle x(t),y(t)\rangle.
\end{aligned} \tag{15}$$

This completes the proof. □

Next, we introduce the fractional Fourier transform as a generalization of the classical Fourier transform.

2.3 Fractional Fourier transform

It is well known that when one performs the FT two times, the time-reverse operation is obtained. When one performs the FT three times, the inverse FT is obtained. Furthermore, performing the FT four times is equivalent to performing an identity operation. Now, one may think what will be obtained when the FT is performed a non-integer number of times The fractional Fourier transform (FRFT) can be viewed as performing the FT $\{2\alpha/\pi\}$ times, where $\{2\alpha/\pi\}$ can be a non-integer value. The fractional Fourier transform (FRFT) has played an important role in signal processing [7] optics [8, 9], image processing [10], and quantum mechanics [11]. As a generalization of the conventional Fourier transform (FT), the FRFT implements an order parameter which acts on the conventional Fourier transform operator and can process time-varying signals and non-stationary signals. With variation of the fractional parameter, the FRFT transforms the signal into the fractional Fourier domain representation, which is oriented by corresponding rotation angle with respect to the time axis in the counter-clockwise direction. Using a global kernel, the FRFT shows the overall fractional Fourier domain contents. Hence, the time-frequency representation should be extended to the time-fractional Fourier frequency domain. Let us define fractional Fourier transform.

Definition 3. *Let $x(t)$ be a signal in $L^2(\mathbb{R})$, then the fractional Fourier transform of $x(t)$ is defined as*

$$\mathscr{F}_\alpha[x(t)](\xi) = \int_{\mathbb{R}} K_\alpha(t,\xi)x(t)dt, \tag{16}$$

where α is a angular parameter and $K_\alpha(t,\xi)$ is the kernel of the FRFT and is given by

$$K_\alpha(t,\xi) = \begin{cases} \sqrt{\dfrac{1-i\cot\alpha}{2\pi}}e^{\frac{i}{2}(t^2+\xi^2)\cot\alpha - i\xi t\csc\alpha} & for \quad \alpha \neq n\pi, \\ \delta(t-\xi) & for \quad \alpha = 2n\pi, \\ \delta(t+\xi) & for \quad \alpha = (2n\pm 1)\pi, \quad n\in\mathbb{Z}. \end{cases} \tag{17}$$

and the corresponding inversion formula is also a FRFT with angle $-\alpha$ and is given by

$$x(t) = \mathscr{F}_{-\alpha}\{\mathscr{F}_\alpha[x(t)](\xi)\}(t) = \int_{\mathbb{R}} \mathscr{F}_\alpha[x(t)](\xi)K_{-\alpha}(t,\xi)d\xi. \tag{18}$$

It is easy to see that, when $\alpha = 0, \pi/2, \pi$ and $3\pi/2$, the FRFT is reduced to the identity operation, the FT, time-reverse operation, and the IFT, respectively.

Assuming that $u(t) = e^{it^2\cot\alpha/2}x(t)$, then for $\alpha \neq n\pi$ the FRFT (16) can be rewritten as

$$\mathscr{F}_\alpha[x(t)](\xi) = \sqrt{\frac{1-i\cot\alpha}{2\pi}}e^{i\xi^2\cot\alpha/2}\left(\frac{1}{\sqrt{2\pi}}\int_{\mathbb{R}} e^{-i\xi t\csc\alpha}u(t)\right) \tag{19}$$

$$= \sqrt{\frac{1-i\cot\alpha}{2\pi}}e^{i\xi^2\cot\alpha/2}\mathscr{F}[u](\xi\csc\alpha). \tag{20}$$

It is clear from (20) that the FRFT can be viewed as a chirp-Fourier-chirp transformation.

Next, we highlight some properties of FRFT.

Theorem 7. *Let $x(t), y(t) \in L^2(\mathbb{R})$ and $k, \xi_0 \in \mathbb{R}$, then the FRFT satisfies following properties:*

1. *Translation:* $\mathscr{F}_\alpha[x(t-k)](\xi) = e^{\frac{1}{2}ik^2\cos\alpha\sin\alpha - ik\xi\sin\alpha}\mathscr{F}_\alpha[x(t)](\xi)(\xi - k\cos\alpha)$.

2. *Modulation:* $\mathscr{F}_\alpha\left[e^{i\xi_0 t}x(t)\right](\xi) = e^{i\xi_0\xi\cos\alpha - \frac{i}{2}\xi_0^2\sin\alpha\cos\alpha}\mathscr{F}_\alpha[x(t)](\xi - \xi_0\sin\alpha)$.

3. *Orthogonality Relation:* $\langle\mathscr{F}_\alpha[x(t)], \mathscr{F}_\alpha[y(t)]\rangle = \langle x(t), y(t)\rangle$.

Proof. For the sake of brevity, we omit proof of translation and modulation properties and prove only orthogonality relation.

We have

$$\begin{aligned}\langle \mathscr{F}_\alpha[x(t)], \mathscr{F}_\alpha[y(t)]\rangle &= \int_{\mathbb{R}} \mathscr{F}_\alpha[x(t)](\xi)\overline{\mathscr{F}_\alpha[y(t)](\xi)}d\xi \\ &= \int_{\mathbb{R}}\int_{\mathbb{R}}\int_{\mathbb{R}} K_\alpha(t,\xi)x(t)\overline{K_\alpha(s,\xi)y(s)dsdtd\xi} \\ &= \int_{\mathbb{R}}\int_{\mathbb{R}} x(t)\overline{y(s)}\left(\int_{\mathbb{R}} K_\alpha(t,\xi)\overline{K_\alpha(s,\xi)}d\xi\right)dsdt \\ &= \int_{\mathbb{R}}\int_{\mathbb{R}} x(t)\overline{y(s)}\delta(t-s)dsdt \\ &= \int_{\mathbb{R}} x(t)\overline{y(s)}dt \\ &= \langle x(t), y(t)\rangle.\end{aligned}$$

This completes the proof. □

Since the FRFT is a generalization of the FT, many properties, applications, and operations associated with FT can be generalized by using the FRFT. The FRFT is more flexible than the FT and performs even better in many signal processing and optical system analysis applications.

In the sequel, we introduce linear canonical transform, which is a generalized version of the classical Fourier transform with four parameters.

2.4 Linear canonical transform

The linear canonical transform (LCT) introduced by Moshinsky and Quesne [12] has a total of four parameters. It is not only a generalization of the FT, but also a generation of the FRFT, the scaling operation. As the FRFT, the LCT was first used for solving differential equations and analyzing optical systems. Recently, after the applications of FRFT were developed, the roles of the LCT for signal processing have also been examined. Due to the extra degrees of freedom and simple geometrical manifestation, the LCT is more flexible than other transforms and is as such suitable as well as powerful tool for investigating deep problems in science and engineering [13–16]. Now, we shall define linear canonical transform (LCT).

Definition 4. *Consider the second order matrix* $M_{2\times2} = \begin{bmatrix} a & b \\ c & d \end{bmatrix}$. *Then the linear canonical transform of any* $x(t) \in L^2(\mathbb{R})$ *with respect to the uni-modular matrix* $M_{2\times2} = \begin{bmatrix} a & b \\ c & d \end{bmatrix}$ *is defined by*

$$\mathscr{L}_M[x(t)](\xi) = \begin{cases} \int_{\mathbb{R}} \mathcal{K}_M(t,\xi)x(t)dt & b \neq 0 \\ \sqrt{d}\exp\left\{\frac{cd\xi^2}{2}\right\}f(d\xi) & b = 0. \end{cases} \tag{21}$$

where $\mathcal{K}_M(t,\xi)$ is the kernel of linear canonical transform and is given by

$$\mathcal{K}_M(t,\xi) = \frac{1}{\sqrt{2\pi ib}} e^{\frac{i}{2b}\left(at^2 - 2t\xi + d\xi^2\right)}, \quad b \neq 0. \tag{22}$$

When $b \neq 0$, the inverse LCT is given by

$$f(t) = \mathcal{L}_{M^{-1}}\left\{\mathcal{L}_M[x((t)](\xi)\right\}(t) = \int_{\mathbb{R}} \mathcal{L}^M[x(t)](\xi)\overline{\mathcal{K}_M(t,\xi)}d\xi \tag{23}$$

where the kernel $\overline{\mathcal{K}_M(t,\xi)} = \mathcal{K}_{M^{-1}}(t,\xi)$ and M^{-1} denotes the inverse of matrix M. For typographical convenience we write the matrix $M = (a;b;c;d)$.

By changing the matrix parameter $M = (a;b;c;d)$, the LCT boils down to various integral transforms such as:

- When $M = (0,1,-1,0)$, the LCT turns out to be Fourier transform(FT):

$$\mathcal{L}_M[x(t)] = \sqrt{-i}\mathcal{F}[x(t)].$$

- When $M = (0,-1,1,0)$, the LCT turns out to be inverse Fourier transform(IFT):

$$\mathcal{L}_M[x(t)] = \sqrt{i}\mathcal{F}^{-1}[x(t)].$$

- When $M = (\cos\alpha,\ \sin\alpha, -\sin\alpha,\ \cos\alpha)$, the LCT becomes the FRFT:

$$\mathcal{L}_M[x(t)] = \sqrt{e^{-i\alpha}}\mathcal{F}_\alpha[x(t)].$$

- When $M = \left(\lambda, 0, 0, \frac{1}{\lambda}\right)$, the LCT becomes a scaling operation:

$$\mathcal{L}_M[x(t)] = \sqrt{\frac{1}{\lambda}}x\left(\frac{\xi}{\lambda}\right).$$

- When $M = (1,0,\beta,1)$, the LCT becomes a chirp multiplication operation:

$$\mathcal{L}_M[x(t)] = e^{\frac{i}{2}\beta\xi^2}x(\xi).$$

Moreover Fresnel transform can be viewed with matrix $(1,b,0,1)$ and the Laplace transform can be obtained with $(0,i,i,0)$.

From (21), we have for $b \neq 0$

$$\begin{aligned}\mathcal{L}_M[x(t)](\xi) &= \int_{\mathbb{R}} \mathcal{K}_M(t,\xi)x(t)dt \\ &= \frac{1}{\sqrt{2\pi ib}}\int_{\mathbb{R}} e^{\frac{i}{2b}\left(at^2-2t\xi+d\xi^2\right)}x(t)dt \\ &= \frac{1}{\sqrt{2\pi ib}}e^{\frac{i}{2b}d\xi^2}\int_{\mathbb{R}} e^{-\frac{i}{b}\xi t}\left(x(t)e^{\frac{i}{2b}at^2}\right)dt \\ &= \frac{1}{\sqrt{2\pi ib}}e^{\frac{i}{2b}d\xi^2}\mathcal{F}[g(t)](\xi/b),\end{aligned} \tag{24}$$

where $g(t) = x(t)e^{\frac{i}{2b}at^2}$.

Thus, it is clear from (24), that LCT can be regarded as a chirp-Fourier-chirp transformation.

Next, we investigate some basic properties associated with LCT.

Theorem 8. *Let $x(t), y(t) \in L^2(\mathbb{R})$ and $k, \xi_0 \in \mathbb{R}$, then the LCT satisfies following properties:*

1. *Translation:* $\mathscr{L}_M[x(t-k)](\xi) = e^{ikc\xi - \frac{i}{2}k^2 ac}\mathscr{L}_M[x(t)](\xi - ak)$.

2. *Modulation:* $\mathscr{L}_M\left[e^{i\xi_0 t}x(t)\right](\xi) = e^{id\xi_0\xi - \frac{i}{2}b\xi_0^2}\mathscr{L}_M[x(t)](\xi - b\xi_0)$.

3. *Parity:* $\mathscr{L}_M[x(-t)](\xi) = \mathscr{L}_M[x(t)](-\xi)$.

4. *Orthogonality Relation:* $\langle \mathscr{L}_M[x(t)], \mathscr{L}_M[y(t)]\rangle = \langle x(t), y(t)\rangle$.

Proof. To be specific, we shall only prove the translation property, the rest of the properties follows similarly.

For any real k, we have

$$\begin{aligned}
\mathscr{L}_M[x(t-k)](\xi) &= \int_{\mathbb{R}} \mathcal{K}_M(t,\xi)x(t-k)dt \\
&= \frac{1}{\sqrt{2\pi ib}}\int_{\mathbb{R}} e^{\frac{i}{2b}\left(at^2 - 2t\xi + d\xi^2\right)}x(t)dt \\
&= \frac{1}{\sqrt{2\pi ib}}\int_{\mathbb{R}} e^{\frac{i}{2b}\left(a(s+k)^2 - 2(s+k)\xi + d\xi^2\right)}x(s)ds \\
&= e^{ikc\xi - \frac{i}{2}k^2 ac}\frac{1}{\sqrt{2\pi ib}}\int_{\mathbb{R}} e^{\frac{i}{2b}\left(as^2 - 2s(\xi - ak) + d\left(\xi - ak^2\right)\right)}x(s)ds \\
&= e^{ikc\xi - \frac{i}{2}k^2 ac}\mathscr{L}_M[x(t)](\xi - ak).
\end{aligned}$$

This completes the proof. □

Finally, we will define quadratic-phase Fourier transform.

2.5 Quadratic-phase Fourier transform

The most neoteric generalization of the classical Fourier transform (FT) with five real parameters appeared *via* the theory of reproducing kernels is known as the quadratic-phase Fourier transform (QPFT) [17]. It treats both the stationary and non-stationary signals in a simple and insightful way that are involved in radar, signal processing, and other communication systems [18–25]. Here, we gave the notation and definition of the quadratic-phase Fourier transform and study some of its properties.

Definition 5. *For a real parameter set $\Lambda = (a, b, c, d, e)$ with $b \neq 0$, the quadratic-phase Fourier transform of any signal $f \in L^2(\mathbb{R})$ is defined as*

$$\mathcal{Q}_\Lambda[x(t)](\xi) = \int_{\mathbb{R}} K_\Lambda(t,\xi)x(t)dt, \tag{25}$$

where $k_\Lambda(t,\xi)$ is the kernel signal of the QPFT and is given by

$$K_\Lambda(t,\xi) = \frac{1}{\sqrt{2\pi}}e^{-i\left(at^2 + b\xi t + c\xi^2 + dt + e\xi\right)}, \tag{26}$$

and corresponding inversion formula is given by

$$x(t) = \mathcal{Q}_\Lambda^{-1}(\mathcal{Q}_\Lambda[x(t)](\xi))(t) = \int_{\mathbb{R}} \overline{K_\Lambda(t,\xi)}\, \mathcal{Q}_\Lambda[x(t)](\xi) d\xi. \tag{27}$$

The novel QPFT (5) can be considered as a cluster of several existing integral transforms ranging from the classical Fourier to the much recent special affine Fourier transform. Nevertheless, many signal processing operations, such as scaling,shifting and time reversal, can also be performed *via* the QPFT (5).

Now, we will establish some properties of the quadratic-phase Fourier transform.

Theorem 9. *Let $x(t), y(t) \in L^2(\mathbb{R})$ and $k, \xi_0 \in \mathbb{R}$, then the QPFT satisfies following properties:*

1. *Modulation:* $\mathcal{Q}_\Lambda\left[e^{i\xi_0 t}x(t)\right](\xi) = e^{i\left(c\left(b^{-2}\xi_0^2 - 2b^{-1}\xi\xi_0\right) - eb^{-1}\xi_0\right)} \mathcal{Q}_\lambda[x(t)]\left(\xi - b^{-1}\xi_0\right).$

2. *Parity:* $\mathcal{Q}_\Lambda[x(-t)](\xi) = \mathcal{Q}_{\Lambda'}[x(t)](-\xi)$, *where* $\Lambda' = (a, b, c, -d, -e).$

3. *conjugation:* $\mathcal{Q}_\Lambda\left[\overline{x(t)}\right](\xi) = \overline{\mathcal{Q}_{-\Lambda}[x(t)](\xi)}$, *where* $-\Lambda = (-a, -b, -c, -d, -e).$

4. *Orthogonality Relation:* $\langle \mathcal{Q}_\Lambda[x(t)], \mathcal{Q}_\Lambda[y(t)]\rangle = \frac{1}{b}\langle x(t), y(t)\rangle.$

Proof. For the sake of brevity, we avoid proof. □

Author details

Mohammad Younus Bhat
Department of Mathematical Sciences, Islamic University of Science and Technology, Kashmir, India

*Address all correspondence to: gyounusug@gmail.com

References

[1] Fourier JB. Theorie analytique de la chaleur. Paris: Chez Firmin Didot, Pere et Fils; 1822;499-508

[2] Durak L, Arikan O. Short-time Fourier transform: Two fundamental properties and an optimal implementation. IEEE Transactions on Signal Processing. 2003; **51**(1231):42

[3] Boggess A, Narcowich FJ. A First Course in Wavelets with Fourier Analysis. Upper Saddle River, New Jersey: Prentice Hall; 2001

[4] Bracewell RN. The Fourier Transform and its Applications. Third ed. Boston: McGraw-Hill; 2000

[5] Howell KB. Principles of Fourier Analysis. Boca Raton: Chapman & Hall/ CRC; 2001

[6] Olson T. Applied Fourier Analysis. NewYork: Birkhauser; 2017

[7] Almeida L. The fractional Fourier transform and time-frequency representations. IEEE Transactions on Signal Processing. 2022;**42**(11): 3084-3091

[8] Ozaktas HM, Mendlovic D. Fourier transforms of fractional order and their optical interpretation. Optics Communication. 1993;**101**(3–4):163-169

[9] Ozaktas HM, Zalevsky Z, Kutay M. The Fractional Fourier Transform with Applications in Optics and Signal Processing. New York: Wiley; 2001

[10] Djurovic I, Stankovic S, Pitas I. Digital watermarking in the fractional Fourier transformation domain. Journal of Network and Computer Applications. 2001;**24**(2):167-173

[11] Namias V. The fractional order Fourier transform and its application to quantum mechanics. IMA Journal of Applied Mathematics. 1980;**25**(3): 241-265

[12] Moshinsky M, Quesne C. Linear canonical transformations and their unitary representations. Journal of Mathematical Physics. 1971;**12**(8): 1772-1780

[13] Barshan B, Kutay MA, Ozaktas HM. Optimal filtering with linear canonical transformations. Optics Communication. 1997;**135**:32-36

[14] Healy JJ, Kutay MA, Ozaktas HM, Sheridan JT. Linear Canonical Transforms. New York: Springer; 2016

[15] Hennelly BM, Sheridan JT. Fast numerical algorithm for the linear canonical transform. Journal of the Optical Society of America A. 2005;**22**: 928-937

[16] Pei SC, Ding JJ. Eigen functions of linear canonical transform. IEEE Transactions on Signal Processing. 2002; **50**:11-26

[17] Saitoh S. Theory of reproducing kernels: Applications to approximate solutions of bounded linear operator functions on Hilbert spaces. American Mathematical Society Transformation Series. 2010;**230**(2):107-134

[18] Castro LP, Minh LT, Tuan NM. New convolutions for quadratic-phase Fourier integral operators and their applications. Mediterranean Journal of Mathematics. 2018;**15**:1-17

[19] Castro LP, Haque MR, Murshed MM, Saitoh S, Tuan NM.

Quadratic Fourier transforms. Annals of Functional Analysis AFA. 2014;**5**(1): 10-23

[20] Bhat MY, Dar AH, Urynbassarova D, Urynbassarova A. Quadratic-phase wave packet transform. Optik - International Journal for Light and Electron Optics. 2022;**261**:169120

[21] Sharma PB. The Wigner-Ville distribution associate with quadratic-phase Fourier transform. AIP Conferenc Proceeding. 2022;**2435**(1):020028

[22] Prasad A, Sharma PB. Convolution and product theorems for the quadratic-phase Fourier transform. Georgian Mathematical Journal. 2022;**29**(4): 595-602

[23] Prasad A, Sharma PB. The quadratic-phase Fourier wavelet transform. Mathematicsl Methods in the Applied Sciences. DOI: 10.1002/mma.6018

[24] Bhat MY, Dar AH. Quadratic-phase scaled wigner distribution: Convolution and correlation. SIViP. 2023;**17**:2779-2788

[25] Bhat MY, Dar AH. Quadratic-phase s–transform.e-prime. Advanced Electrical Engineering Electronics Energy. 2023. DOI: 10.1016/j.prime.2023.100162

Chapter 2

Hybrid Transforms

Didar Urynbassarova and Altyn Urynbassarova

Abstract

Hybrid transforms are constructed by associating the Wigner-Ville distribution (WVD) with widely-known signal processing tools, such as fractional Fourier transform, linear canonical transform, offset linear canonical transform (OLCT), and their quaternion-valued versions. We call them hybrid transforms because they combine the advantages of both transforms. Compared to classical transforms, they show better results in applications. The WVD associated with the OLCT (WVD-OLCT) is a class of hybrid transform that generalizes most hybrid transforms. This chapter summarizes research on hybrid transforms by reviewing a computationally efficient type of the WVD-OLCT, which has simplicity in marginal properties compared to WVD-OLCT and WVD.

Keywords: time-frequency analysis, Wigner-Ville distribution, offset linear canonical transform, hybrid transform, linear frequency modulated (LFM) signal

1. Introduction

The linear canonical transform (LCT) [1–4] and its generalization, the offset linear canonical transform (OLCT) [5, 6] are introduced to study non-stationary signals (audio, image, biomedical, linear frequency modulated (LFM) signals). OLCT has five degrees of freedom, and LCT has three degrees of freedom, which makes them more flexible than the well-known fractional Fourier transform (FrFT) [7] with one degree of freedom and the Fourier transform (FT) with no freedom. Various applications of LCT have been found in the different fields of optics and signal processing. In fact, the properties and applications of the OLCT are similar to the LCT, but they are more general than the LCT, thanks to its two extra parameters, which correspond to time-shift and frequency modulation. It is proven that the Wigner-Ville distribution (WVD) plays a major role in time-frequency signal analysis and processing.

The LFM signal is used in communications, radar and sonar systems. Consequently, LFM signal detection and estimation is one of the most important topics in engineering. The WVD and LCT/OLCT are used in LFM signal processing, but they have their disadvantages:

- WVD does not fully exploit the phase feature of LFM signal;
- LCT/OLCT cannot gather signal energy strongly like WVD.

This results in poor performance under a low signal-to-noise ratio for detection and estimation. Recently, for the purpose to improve the performance of LFM signal detection and estimation, several researchers have associated WVD with the FrFT, LCT, and OLCT, respectively [8–27]. Results show that such transforms exploit the advantages of both transforms, which is why we call them hybrid transforms. The aim of this chapter is to review and summarize research on hybrid transforms by studying WVD association with the OLCT (WVD-OLCT) definitions and properties.

2. Preliminaries

2.1 Wigner-Ville distribution

FT analysis originated long ago and is used in many areas of mathematics and engineering, including quantum mechanics, wave propagation, turbulence, signal analysis and processing. In spite of remarkable success, the FT analysis seems to be inadequate for studying some problems for the following reasons:

- There is no local information in the FT analysis since it does not reflect the change of frequency with time;
- The FT analysis investigates problems either in the time domain or in the frequency domain, but not simultaneously in both domains.

Therefore, we see that FT is sufficient to study signals that are statistically invariant over time, e.g. stationary signals. Naturally, we are surrounded by many signals: audio, video, radar, biomedical signals, etc., all those signals are non-stationary. FT is insufficient to do a complete analysis for such signals because it requires both time-frequency representations of the signal. So it was necessary to define a single transformation of time and frequency domains.

Historically, Eugene Paul Wigner, the 1963 Nobel Prize winner in physics, in 1932 first introduced a fundamental nonlinear transformation to study quantum corrections for classical statistical mechanics in the form [28].

$$\mathcal{W}_{\psi}(x,p)=\frac{1}{h}\int_{\mathbb{R}}\psi\left(x-\frac{\tau}{2}\right)\overline{\psi}\left(x+\frac{\tau}{2}\right)\exp\left(\frac{ip\tau}{\hbar}\right)d\tau, \tag{1}$$

where the wave function $\psi(x)$ satisfies the one-dimensional Schrödinger equation, the quantum mechanical position x and momentum p are independent variables, and $h=2\pi\hbar$ is the Planck constant. The Wigner distribution $\mathcal{W}_{\psi}(x,p)$ has many important properties and is found to behave as a distribution function defined on a phase space consisting of points (x,p). The most remarkable properties of the Wigner distribution include the marginal integrals in the position and momentum domains as follows [29, 30].

$$\begin{aligned}\int_{\mathbb{R}}\mathcal{W}_{\psi}(x,p)dx&=|\varphi(p)|^2,\\ \int_{\mathbb{R}}\mathcal{W}_{\psi}(x,p)dp&=|\psi(x)|^2,\end{aligned} \tag{2}$$

and the total energy of the wave function ψ in the (x, p) space

$$\int_{\mathbb{R}^2} \mathcal{W}_\psi(x, p)dxdp = \int_{\mathbb{R}} |\psi(x)|^2 dx = \|\psi\|. \tag{3}$$

In the context of non-stationary signal analysis, in 1948 Jean-Andre Ville independently re-derived the Wigner distribution given in Eq. (1) as a quadratic representation of the local time-frequency energy of a signal [31]. Besides linear time-frequency representations of a signal like the Gabor transform, the Zak transform, and the short-time Fourier transform, the WVD (or Wigner-Ville transform (WVT)) occupies a central position in the field of quadratic time-frequency representations and it is recognized as a valuable method/tool for time-frequency of time-varying signals and non-stationary random processes.

With its remarkable structure and properties, the WVD has been regarded as the main distribution of all the time-frequency distributions and used as the classical and fundamental time-frequency analysis tool in different areas of physics and engineering. Particularly, it has been used for instantaneous frequency estimation, spectral analysis of random signals, detection and classification, algorithms for computer implementation, and has a wide range of applications in vision, X-ray diffraction of crystals, pattern recognition, radar, and sonar. Additionally, it has been applied to the analysis of seismic data, speech, and phase distortions in audio engineering problems.

Definition 1 (WVD). If f belong to the Hilbert space $L^2(\mathbb{R})$, the WVD $\mathcal{W}_f$ of signal f is defined as [3, 29, 30].

$$\mathcal{W}_f(t, u) = \int_{\mathbb{R}} f\left(t + \frac{\tau}{2}\right)\overline{f\left(t - \frac{\tau}{2}\right)}e^{-iu\tau}d\tau. \tag{4}$$

It is easy to see that the WVD is the FT of the instantaneous autocorrelation function

$$R_f(t, \tau) = f\left(t + \frac{\tau}{2}\right)\overline{f\left(t - \frac{\tau}{2}\right)} \tag{5}$$

with respect to τ.

Some main properties of WVD are summarized in **Table 1**. For some recent works and surveys on the WVD, we refer readers to [3, 29, 30] and the references therein.

2.2 Linear canonical transform

The LCT is a four-parameter (a, b, c, d) integral transform that was introduced in the 1970s by Collins, and Moshinsky and Quesne to analyze optical systems and solve differential equations [1, 2]. After the fast algorithm for calculating the discrete LCT was proposed in [32], the LCT was widely used to process non-stationary signals. It has been applied in radar system analysis, filter design, watermarking, phase retrieval, pattern recognition, signal synthesis, and in other areas of engineering sciences. With intensive research, many properties of the LCT are well studied. Transforms and operations, such as the FT, FrFT, Fresnel transform (FRST), Laplace transform, fractional Laplace transform, time scaling, and chirp operations are the special cases of the LCT.

Property	Formulation
Conjugation symmetry	$\mathcal{W}_f(t, u) = \overline{\mathcal{W}_f(t, u)}$
Time shifting (Translation)	$\mathcal{W}_{f'}(t, u) = \mathcal{W}_f(t - \lambda, u), f'(t) = f(t - \lambda)$
Frequency shifting (Modulation)	$\mathcal{W}_{f'}(t, u) = \mathcal{W}_f(t, u - u_0), f'(t) = f(t)e^{iu_0 t}$
Time marginal	$\int_{\mathbb{R}} \mathcal{W}_f(t, u)du = \lvert f(t)\rvert^2$
Frequency marginal	$\int_{\mathbb{R}} \mathcal{W}_f(t, u)dt = \left\lvert \hat{f}(u)\right\rvert^2$
Energy distribution	$\int_{\mathbb{R}^2} \mathcal{W}_f(t, u)dtdu = \int_{\mathbb{R}} \lvert f(t)\rvert^2 dt = \langle f(t), f(t)\rangle$
Moyal's formula	$\int_{\mathbb{R}^2} \mathcal{W}_f(t, u)\overline{[\mathcal{W}_g(t, u)]}dtdu = \lvert\langle f, g\rangle\rvert^2$

Table 1.
Properties of the WVD.

In some works, the LCT is known under different names as the Collins formula, Moshinsky and Quesne integrals, extended fractional Fourier transform, quadratic-phase integral or quadratic-phase system, generalized Fresnel transform, generalized Huygens integral [33], ABCD transform [34], and affine Fourier transform [35], etc.

Definition 2 (LCT). The LCT $\mathcal{L}_{\mathbf{A}}$ of a signal $f(t)$ with matrix $\mathbf{A} = (a, b, c, d)$, where $a,b,c,d \in \mathbb{R}$ are real parameters and $\det(\mathbf{A}) = ad - bc = 1$, is defined as [2–4]

$$\mathcal{L}_{\mathbf{A}}\{f(t)\}(u) = \begin{cases} \int_{\mathbb{R}} f(t) \dfrac{1}{\sqrt{i2\pi b}} e^{i\left(\frac{a}{2b}t^2 - \frac{1}{b}tu + \frac{d}{2b}u^2\right)} dt, \, b \neq 0, \\ \sqrt{d} e^{\frac{i}{2}cdu^2} f(du), \, b = 0. \end{cases} \tag{6}$$

From the definition of LCT, we can see that, when the parameter $b = 0$, the LCT is a scaling transformation coupled with amplitude and quadratic phase modulation and it is of no particular interest to our object. Therefore, without loss of generality, in this chapter we always assume $b \neq 0$.

A detailed and comprehensive view of LCT can be found in [2, 3] and the references therein.

2.3 Offset linear canonical transform

The OLCT is a six-parameter $(a, b, c, d, u_0, \omega_0)$ integral transform, which has been shown as a powerful tool and received much attention in signal processing and optics. It is a time-shifted and frequency-modulated version of the LCT. In some works OLCT called the special affine Fourier transform [35–37] and the inhomogeneous canonical transform [38].

Definition 3 (OLCT). The OLCT $\mathcal{O}_{\mathbf{A}}$ of a signal $f(t)$ with real parameters of matrix $\mathbf{A} = (a, b, c, d, u_0, \omega_0)$, where $a,b,c,d,u_0,\omega_0 \in \mathbb{R}$ are real parameters and $\det(\mathbf{A}) = 1$, is defined as [6, 19]

$$\mathcal{O}_{\mathbf{A}}\{f(t)\}(u) = \begin{cases} \int_{\mathbb{R}} f(t) K_{\mathbf{A}}(t, u) dt, \, b \neq 0, \\ \sqrt{d} e^{i\frac{cd}{2}(u-u_0)^2 + j\omega_0 u} f(d(u - u_0)), \quad b = 0. \end{cases} \tag{7}$$

where $K_{\mathbf{A}}(t,u)$ is the OLCT kernel and expressed as

$$K_{\mathbf{A}}(t,u) = \frac{1}{\sqrt{i2\pi b}} e^{i\left(\frac{a}{2b}t^2 - \frac{1}{b}t(u-u_0) + \frac{d}{2b}\left(u^2+u_0^2\right) - \frac{u}{b}(du_0 - b\omega_0)\right)}. \tag{8}$$

From Eq. (7) it can be seen that for case $b = 0$ the OLCT is simply a time scaled version of f multiplied by a linear chirp. Therefore, from now we restrict our attention to OLCT for the case $b \neq 0$. And without loss of generality, we assume $b > 0$ in the following sections of this chapter.

A number of widely known classical transforms and mathematical operations related to signal processing and optics are special cases of the OLCT. The OLCT converts to its special cases when taking different parameters of matrix $\mathbf{A}$. For example, the OLCT with parameters $(a,b,c,d,u_0,\omega_0) = (a,b,c,d,0,0)$ reduces to LCT; when $\mathbf{A} = (\cos\theta, \sin\theta, -\sin\theta, \cos\theta, 0, 0)$, it becomes the FrFT; when $\mathbf{A} = (0,1,-1,0,0,0)$, the OLCT becomes FT; when $\mathbf{A} = (1,b,0,1,0,0)$, it becomes FRST; and when $\mathbf{A} = \left(d^{-1},0,0,d,0,0\right)$, it becomes time scaling operation. Multiplication by Gaussian or chirp function is obtained with an $\mathbf{A} = (1,0,\tau,1,0,0)$ [1]. The offset Fourier transform $\mathbf{A} = (0,1,-1,0,u_0,\omega_0)$, offset fractional Fourier transform $\mathbf{A} = (\cos\theta, \sin\theta, -\sin\theta, \cos\theta, u_0, \omega_0)$, frequency modulation $\mathbf{A} = (1,0,0,1,0,\omega_0)$, and time shifting $\mathbf{A} = (1,0,0,1,u_0,0)$ are also special cases of the OLCT. The OLCT is able to extend their properties and applications and can solve some problems that cannot be solved well by these operations. In fact, offset versions of FT, FrFT, and LCT are similar to the classical FT, FrFT, and LCT, but they are more flexible than the classical ones, and mainly useful for analyzing optical systems with prisms or shifted lenses. The OLCT has a close relationship with its special cases. So it is practically useful to develop relevant theorems for OLCT. By developing theories for OLCT, we can gain a deeper understanding of its special cases and transfer knowledge from one subject to another. As a generalization of many other linear transforms, the OLCT has found wide applications in applied mathematics, signal processing, and optical system modeling [5, 6, 19, 34, 35, 37].

2.4 Previous results

With the development of the FrFT, Lohmann in [8] and Almeida in [9] investigated the relationship between the WVD and the FrFT. They show that the WVD of the FrFTed signal can be seen as a rotation of the WVD in the time-frequency plane. In this direction, based on the properties of the FrFT, the LCT, and the WVD, Pei and Ding [10] investigated and discussed the relations between the common fractional and canonical operators. The WVD associated with the LCT, named LCWD, denoted as $\mathcal{WD}_{\mathbf{A}}$, given in [10] is useful for the separation of multi-component signals. It is defined as [10, 18].

$$\mathcal{WD}_{\mathbf{A}}(u,v) = \int_{\mathbb{R}} \mathcal{L}_{\mathbf{A}}\left(u+\frac{\tau}{2}\right)\overline{\mathcal{L}_{\mathbf{A}}\left(u-\frac{\tau}{2}\right)} e^{-iv\tau} d\tau, \tag{9}$$

where $\mathcal{L}_{\mathbf{A}}(u)$ is the LCT of signal $f(t)$ with parameter matrix $\mathbf{A} = (a,b,c,d)$.

Unlike the definition of LCWD, Bai et al. obtained generalized type of WVD in the LCT domain, named WVD-LCT (or WDL), denoted as $\mathcal{WDL}_f$, by substituting FT kernel $e^{-iu\tau}$ with LCT kernel $\frac{1}{\sqrt{i2\pi b}} e^{i\left(\frac{a}{2b}\tau^2 - \frac{1}{b}\tau u + \frac{d}{2b}u^2\right)}$ [11].

$$\mathcal{WDL}_f(t,u) = \int_{\mathbb{R}} f\left(t+\frac{\tau}{2}\right)\overline{f\left(t-\frac{\tau}{2}\right)} \frac{1}{\sqrt{i2\pi b}} e^{i\left(\frac{a}{2b}\tau^2 - \frac{1}{b}\tau u + \frac{d}{2b}u^2\right)} d\tau. \tag{10}$$

The WVD-LCT generalizes the LCWD and WVD. It is easy to see that the WVD-LCT is the LCT of the instantaneous autocorrelation function $R_f(t,\ \tau)$ with respect to τ

$$\mathcal{WDL}_f(t,u) = \mathcal{L}_{\mathbf{A}}\{R_f(t,\ \tau)\}. \tag{11}$$

Also, in [11] authors derived the main properties and applications of the WVD-LCT in the LFM signal detection. Uncertainty principles for the WVD-LCT were studied in [13, 25]. Song et al. presented WVD-LCT applications for quadratic frequency modulated signal parameter estimation in [14]. Convolution and correlation theorems for WVD-LCT are obtained in [16]. In [26] authors proposed a new method of instantaneous frequency estimation by associating the WVD with the LCT, which has a higher capacity for anti-noise and a higher estimation accuracy than WVD. Zhang unified LCWD and WVD-LCT [20], and then presented its special cases with less parameters [21, 22]. Urynbassarova et al. presented the WVD associated with the instantaneous autocorrelation function in the LCT domain, named WL, which has elegance and simplicity in marginal properties and affine transformation relationships compared to the WVD [17]. Similar to this in [27] Xin and Li proposed a new definition of WVD associated with LCT, and its integration form, which estimates two phase coefficients of LFM signal simultaneously and effectively suppresses cross terms for multi-component LFM signal. In [19] introduced the WVD association with the OLCT (WVD-OLCT), which is a generalization of the WVD-LCT and its special cases. Recently, in order to study higher dimensions, WVD associations with the quaternion LCT/OLCT were studied in [39–42], and WVD in the framework of octonion LCT was proposed by Dar and Bhat [43].

3. Definition

The WVD given in Eq. (4) can be re-written as

$$\mathcal{W}_f(t,u) = \int_{\mathbb{R}} f_{\mathcal{F}}\left(t+\frac{\tau}{2}\right)\overline{f_{\mathcal{F}}\left(t-\frac{\tau}{2}\right)} d\tau, \tag{12}$$

where $f_{\mathcal{F}}$ equals to $f(t)$ multiplied with FT kernel e^{-iut}. By substituting FT kernel e^{-iut} with OLCT kernel (Eq. (8)), we will get the following definition of the WVD in the OLCT domain, named WOL, denoted as $\mathcal{WOL}_f$, which is the type of the WVD-OLCT.

Definition 4 (WOL). The WOL $\mathcal{WOL}_f$ of signal f for the parameter matrix $\mathbf{A} = (a, b, c, d, u_0, \omega_0)$ is defined as follows [18]

$$\mathcal{WOL}_f(t,u) = \frac{1}{2\pi|b|} \int_{\mathbb{R}} f\left(t+\frac{\tau}{2}\right)\overline{f\left(t-\frac{\tau}{2}\right)} e^{\frac{ia}{b}\tau t} e^{\frac{i}{b}\tau(u_0-u)} d\tau.$$

The WOL is reduced to the WL, when $\mathbf{A} = (a, b, c, d, 0, 0)$,

$$\mathcal{WOL}_f^{(a,\ b,\ c,\ d,\ 0,\ 0)}(t,u) = \mathcal{WL}_f(t,u). \tag{13}$$

Obviously, when the parameter matrix has the special form $\mathbf{A} = (0, 1, -1, 0, 0, 0)$, the WOL is reduced to the WVD

$$\mathcal{WOL}_f^{(0,\,1,\,-1,\,0,\,0,\,0)}(t, u) = \mathcal{W}_f(t, u). \tag{14}$$

It is clear from Eq. (13) and Eq. (14) that the WOL is a generalization of the WL and the WVD.

4. Properties

Bellow we list some basic properties of the WOL.

Conjugation symmetry property.

The conjugation symmetry property of the WOL is expressed as

$$\mathcal{WOL}_f(t, u) = \overline{\mathcal{WOL}_f(t, u)}. \tag{15}$$

Proof. From the Definition 4, we have

$$\begin{aligned} \overline{\mathcal{WOL}_f(t, u)} &= \overline{\int_{\mathbb{R}} f\left(t+\frac{\tau}{2}\right)\overline{f\left(t-\frac{\tau}{2}\right)} e^{\frac{ia}{b}\tau t} e^{\frac{i}{b}\tau(u_0-u)} d\tau} \\ &= \int_{\mathbb{R}} \overline{f\left(t+\frac{\tau}{2}\right)} f\left(t-\frac{\tau}{2}\right) e^{\frac{ia}{b}(-\tau)t} e^{\frac{i}{b}(-\tau)(u_0-u)} d\tau, \end{aligned} \tag{16}$$

let $-\tau = \tau'$, then we will arrive at

$$\begin{aligned} \overline{\mathcal{WOL}_f(t, u)} &= \int_{\mathbb{R}} f\left(t+\frac{\tau'}{2}\right)\overline{f\left(t-\frac{\tau'}{2}\right)} e^{\frac{ia}{b}\tau' t} e^{\frac{i}{b}\tau'(u_0-u)} d\tau' \\ &= \mathcal{WOL}_f(t, u). \blacksquare \end{aligned} \tag{17}$$

This property shows that the WOL is always a real number.

Time marginal property.

The time marginal property of the WOL is given as

$$\int_{\mathbb{R}} \mathcal{WOL}_f(t, u) du = |f(t)|^2. \tag{18}$$

Proof.

$$\begin{aligned} \int_{\mathbb{R}} \mathcal{WOL}_f(t, u) du &= \frac{1}{2\pi|b|} \int_{\mathbb{R}^2} f\left(t+\frac{\tau}{2}\right)\overline{f\left(t-\frac{\tau}{2}\right)} e^{\frac{ia}{b}\tau t} e^{\frac{i}{b}\tau(u_0-u)} d\tau du \\ &= \frac{1}{2\pi|b|} \int_{\mathbb{R}} f\left(t+\frac{\tau}{2}\right)\overline{f\left(t-\frac{\tau}{2}\right)} e^{\frac{ia}{b}\tau t} e^{\frac{i}{b}u_0\tau} \left(\int_{\mathbb{R}} e^{-\frac{i}{b}u\tau} du\right) d\tau \\ &= \int_{\mathbb{R}} f\left(t+\frac{\tau}{2}\right)\overline{f\left(t-\frac{\tau}{2}\right)} e^{\frac{ia}{2b}\tau^2} e^{\frac{i}{b}u_0\tau} \delta(\tau) d\tau \\ &= |f(t)|^2. \ \blacksquare \end{aligned} \tag{19}$$

Frequency marginal property.
The frequency marginal property of the WOL is given by

$$\int_{\mathbb{R}} \mathcal{WOL}_f(t,u)dt = \left|\hat{f}(u)\right|^2. \tag{20}$$

Proof.

$$\begin{aligned}
\int_{\mathbb{R}} \mathcal{WOL}_f(t,u)dt &= \frac{1}{2\pi|b|}\int_{\mathbb{R}^2} f\left(t+\frac{\tau}{2}\right)\overline{f\left(t-\frac{\tau}{2}\right)} e^{\frac{ia}{b}\tau t} e^{-\frac{i}{b}(u-u_0)\tau} d\tau dt \\
&= \frac{1}{2\pi|b|}\int_{\mathbb{R}^2} f\left(t+\frac{\tau}{2}\right)\overline{f\left(t-\frac{\tau}{2}\right)} e^{\frac{ia}{2b}\left(t^2+t\tau+\frac{\tau^2}{4}-t^2+t\tau-\frac{\tau^2}{4}\right)} e^{-\frac{i}{b}(u-u_0)\left(t+\frac{\tau}{2}+\frac{\tau}{2}-t\right)} d\tau dt \\
&= \frac{1}{2\pi|b|}\int_{\mathbb{R}^2} f\left(t+\frac{\tau}{2}\right)\overline{f\left(t-\frac{\tau}{2}\right)} e^{\frac{ia}{2b}\left(t+\frac{\tau}{2}\right)^2} e^{-\frac{ia}{2b}\left(t-\frac{\tau}{2}\right)^2} e^{-\frac{i}{b}(u-u_0)\left(t+\frac{\tau}{2}+\frac{\tau}{2}-t\right)} d\tau dt.
\end{aligned} \tag{21}$$

Let $\omega = t+\frac{\tau}{2}$ and let $\upsilon = t-\frac{\tau}{2}$, then above equation reduces to the final result

$$\begin{aligned}
\int_{\mathbb{R}} \mathcal{WOL}_f(t,u)dt &= \frac{1}{2\pi|b|}\int_{\mathbb{R}^2} f(\omega)\overline{f(\upsilon)} e^{\frac{ia}{2b}\omega^2} e^{-\frac{ia}{2b}\upsilon^2} e^{\frac{i}{b}u_0(\omega-\upsilon)} e^{-\frac{i}{b}u(\omega-\upsilon)} d\omega d\upsilon \\
&= \left|\hat{f}(u)\right|^2. \blacksquare
\end{aligned} \tag{22}$$

Energy distribution property.
The energy distribuition property of the WOL is given as

$$\int_{\mathbb{R}^2} \mathcal{WOL}_f(t,u)dtdu = \int_{\mathbb{R}} |f(t)|^2 dt. \tag{23}$$

Proof.

$$\begin{aligned}
\int_{\mathbb{R}^2} \mathcal{WOL}_f(t,u)dtdu &= \frac{1}{2\pi|b|}\int_{\mathbb{R}^3} f\left(t+\frac{\tau}{2}\right)\overline{f\left(t-\frac{\tau}{2}\right)} e^{\frac{ia}{b}\tau t} e^{-\frac{i}{b}(u-u_0)\tau} d\tau dt du \\
&= \frac{1}{2\pi|b|}\int_{\mathbb{R}^2} f\left(t+\frac{\tau}{2}\right)\overline{f\left(t-\frac{\tau}{2}\right)} e^{\frac{ia}{b}\tau t} e^{\frac{i}{b}u_0\tau} \left(\int_{\mathbb{R}} e^{-\frac{i}{b}u\tau} du\right) d\tau dt \\
&= \int_{\mathbb{R}} |f(t)|^2 dt. \blacksquare
\end{aligned} \tag{24}$$

Moyal's formula.
The Moyal's formula of the WOL is presented as

$$\int_{\mathbb{R}^2} \mathcal{WOL}_f(t,u)\overline{\mathcal{WOL}_g(t,u)}dtdu = |\langle f, g\rangle|^2. \tag{25}$$

Proof.

$$\begin{aligned}
&\int_{\mathbb{R}^2} \mathcal{WOL}_f(t,u)\overline{\mathcal{WOL}_g(t,u)}dtdu = \\
&= \frac{1}{2\pi|b|}\int_{\mathbb{R}^4} f\left(t+\frac{\tau}{2}\right)\overline{f\left(t-\frac{\tau}{2}\right)}e^{\frac{ia}{b}\tau t}e^{\frac{i}{b}u_0\tau}e^{-\frac{i}{b}u\tau} \\
&\times \frac{1}{2\pi|b|}\overline{g\left(t+\frac{\tau'}{2}\right)}g\left(t-\frac{\tau'}{2}\right)e^{-\frac{ia}{b}\tau' t}e^{-\frac{i}{b}u_0\tau}e^{\frac{i}{b}u\tau'}d\tau d\tau' dtdu \\
&= \frac{1}{2\pi|b|}\int_{\mathbb{R}^3} f\left(t+\frac{\tau}{2}\right)\overline{f\left(t-\frac{\tau}{2}\right)}e^{\frac{ia}{b}\tau t}e^{\frac{i}{b}u_0\tau}e^{-\frac{i}{b}u\tau}d\tau \\
&\times \frac{1}{2\pi|b|}\int_{\mathbb{R}}\overline{g\left(t+\frac{\tau'}{2}\right)}g\left(t-\frac{\tau'}{2}\right)e^{-\frac{ia}{b}\tau' t}e^{-\frac{i}{b}u_0\tau}e^{\frac{i}{b}u\tau'}d\tau' dtdu \\
&= \frac{1}{2\pi|b|}\int_{\mathbb{R}^2} f\left(t+\frac{\tau}{2}\right)\overline{f\left(t-\frac{\tau}{2}\right)}e^{\frac{ia}{b}\tau t}e^{\frac{i}{b}u_0\tau}d\tau \\
&\times \int_{\mathbb{R}}\overline{g\left(t+\frac{\tau'}{2}\right)}g\left(t-\frac{\tau'}{2}\right)e^{-\frac{ia}{b}\tau' t}e^{-\frac{i}{b}u_0\tau}d\tau' dt\frac{1}{2\pi|b|}\int_{\mathbb{R}} e^{\frac{i}{b}u(\tau'-\tau)}du \\
&= \frac{1}{2\pi|b|}\int_{\mathbb{R}^2} f\left(t+\frac{\tau}{2}\right)\overline{f\left(t-\frac{\tau}{2}\right)}e^{\frac{ia}{b}\tau t}e^{\frac{i}{b}u_0\tau}d\tau\int_{\mathbb{R}}\overline{g\left(t+\frac{\tau'}{2}\right)}g\left(t-\frac{\tau'}{2}\right)e^{-\frac{ia}{b}\tau' t}e^{-\frac{i}{b}u_0\tau}\delta(\tau-\tau')d\tau' dt \\
&= \frac{1}{2\pi|b|}\int_{\mathbb{R}}\left[\int_{\mathbb{R}} f\left(t+\frac{\tau}{2}\right)\overline{f\left(t-\frac{\tau}{2}\right)}\overline{g\left(t+\frac{\tau}{2}\right)}g\left(t-\frac{\tau}{2}\right)dt\right]d\tau.
\end{aligned} \tag{26}$$

Now, we make the change of variable $\mu = t - \frac{\tau}{2}$, and come to

$$\begin{aligned}
\int_{\mathbb{R}^2} \mathcal{WOL}_f(t,u)\overline{\mathcal{WOL}_g(t,u)}dtdu &= \frac{1}{2\pi|b|}\int_{\mathbb{R}} f(\mu+\tau)\overline{g(\mu+\tau)}d\tau\left[\overline{\int_{\mathbb{R}} f(\mu)\overline{g(\mu)}d\mu}\right] \\
&= \frac{1}{2\pi|b|}|\langle f,\ g\rangle|^2. \blacksquare
\end{aligned} \tag{27}$$

Property	Formulation
Conjugation symmetry	$\mathcal{WOL}_f(t,u) = \overline{\mathcal{WOL}_f(t,u)}$
Time shifting	$\mathcal{WOL}_{f'}(t,u) = \mathcal{WOL}_f(t-\lambda, u-a\lambda),\ f'(t) = f(t-\lambda)$
Frequency shifting	$\mathcal{WOL}_{f'}(t,u) = \mathcal{WOL}_f(t, u-u_1 b),\ f'(t) = f(t)e^{iu_1 t}$
Time marginal	$\int_{\mathbb{R}} \mathcal{WOL}_f(t,u)du = \lvert f(t)\rvert^2$
Frequency marginal	$\int_{\mathbb{R}} \mathcal{WOL}_f(t,u)dt = \left\lvert\hat{f}(u)\right\rvert^2$
Energy distribution	$\int_{\mathbb{R}^2} \mathcal{WOL}_f(t,u)dtdu = \int_{\mathbb{R}} \lvert f(t)\rvert^2 dt$
Moyal's formula	$\int_{\mathbb{R}^2} \mathcal{WOL}_f(t,u)\left[\overline{\mathcal{WOL}_g(t,\ u)}\right]dtdu = \frac{1}{2\pi\lvert b\rvert}\lvert\langle f,\ g\rangle\rvert^2$

Table 2.
Properties of the WOL.

Some main properties of WOL are summarized in **Table 2**. The comprehensive view on the WOL can be seen in [17, 18].

5. Conclusion

In this chapter, we thoroughly revised research on hybrid transforms, which are constructed by associating WVD with well-known signal processing tools, such as FrFT, LCT, and OLCT. The WVD-OLCT generalizes most hybrid transforms, and the WOL is its special type. It is proven that hybrid transforms have better output in detection and estimation applications. Since the idea of associating two transforms is novel, it needs deep theoretical analysis and lacks diverse applications. Interested readers can develop hybrid transforms into quaternion and octonion algebra. These studies may be helpful in color image processing.

Acknowledgements

This research was funded by the Science Committee of the Ministry of Education and Science of the Republic of Kazakhstan (Grant No. AP14871252).

Author details

Didar Urynbassarova[1*] and Altyn Urynbassarova[2,3]

1 National Engineering Academy of the Republic of Kazakhstan, Almaty, Kazakhstan

2 Department of Technology and Ecology, School of Society, Technology and Ecology, Narxoz University, Almaty, Kazakhstan

3 Faculty of Information Technology, Department of Information Security, L.N. Gumilyov Eurasian National University, Astana, Kazakhstan

*Address all correspondence to: didaruzh@mail.ru

References

[1] Collins SA. Lens-system diffraction integral written in term of matrix optics. Journal of the Optical Society of America. 1970;**60**:1168-1177. DOI: 10.1364/JOSA.60.001168

[2] Moshinsky M, Quesne C. Linear canonical transform and their unitary representations. Journal of Mathematical Physics. 1971;**12**:1772-1783. DOI: 10.1063/1.1665805

[3] Xu TZ, Li BZ. Linear Canonical Transform and its Application. Beijing: Science Press; 2013

[4] Healy JJ, Kutay MA, Ozaktas HM, Sheridan JT, editors. Linear Canonical Transforms: Theory and Applications. 1st ed. New York, NY: Springer; 2016. DOI: 10.1007/978-1-4939-3028-9_1

[5] Stern A. Sampling of compact signals in the offset linear canonical domain. Signal Image Video Process. 2007;**1**: 359-367. DOI: 10.1007/s11760-007-0029-0

[6] Xiang Q, Qin KY. Convolution, correlation, and sampling theorems for the offset linear canonical transform. Signal Image Video Process. 2012;**2014**:433-442. DOI: 10.1007/s11760-012-0342-0

[7] Tao R, Deng B, Wang Y. Fractional Fourier Transform and its Applications. Beijing: Tsinghua University Press; 2009

[8] Lohmann AW. Image rotation, Wigner rotation and the fractional Fourier transform. Journal of the Optical Society of America A. 1993;**10**:2181-2186. DOI: 10.1364/JOSAA.10.002181

[9] Almeida LB. The fractional Fourier transform and time-frequency representations. IEEE Transactions on Signal Processing. 1994;**42**:3084-3091. DOI: 10.1109/78.330368

[10] Pei SC, Ding JJ. Relations between fractional operations and time-frequency distributions, and their applications. IEEE Transactions on Signal Processing. 2001;**49**:1638-1655. DOI: 10.1109/ 78.934134

[11] Bai RF, Li BZ, Cheng QY. Wigner-Ville distribution associated with the linear canonical transform. Journal of Applied Mathematics. 2012;**2012**:1-14. DOI: 10.1155/2012/740161

[12] Yan JP, Li BZ, Chen YH, Cheng QY. Wigner distribution moments associated with the linear canonical transform. International Journal of Electronics. 2013;**100**:473-481. DOI: 10.1080/ 00207217.2012.713018

[13] Li YG, Li BZ, Sun HF. Uncertainty principles for Wigner-Ville distribution associated with the linear canonical transforms. Abstract and Applied Analysis. 2014;**2014**:1-9. DOI: 10.1155/ 2014/470459

[14] Song YE, Zhang XY, Shang CH, Bu HX, Wang XY. The Wigner-Ville distribution based on the linear canonical transform and its applications for QFM signal parameters estimation. Journal of Applied Mathematics. 2014;**2014**:1-8. DOI: 10.1155/2014/516457

[15] Wei DY, Li Y. Linear canonical Wigner distribution and its application. Optik. 2014;**125**:89-92. DOI: 10.1016/j. ijleo.2013.07.007

[16] Bahri M, Ashino R. Convolution and correlation theorems for Wigner-Ville distribution associated with linear canonical transform. In: 12th International Conference on Information

Technology–New Generations; 13-15 April 2015. pp. 341-346. DOI: 10.1109/ ITNG.2015.61

[17] Urynbassarova D, Li BZ, Tao R. The Wigner-Ville distribution in the linear canonical transform domain. IAENG International Journal of Applied Mathematics. 2016;**46**:559-563

[18] Urynbassarova D, Urynbassarova A, Al-Hussam E. The Wigner-Ville distribution based on the offset linear canonical transform domain. In: 2nd International Conference on Modeling, Simulation and Applied Mathematics (MSAM2017); Advances in Intelligent Systems Research. March 2017. pp 139-142. DOI: 10.2991/msam-17.2017.31

[19] Urynbassarova D, Li BZ, Tao R. Convolution and correlation theorems for Wigner-Ville distribution associated with the offset linear canonical transform. Optik. 2017;**157**:455-466. DOI: 10.1016/j.ijleo.2017.08.099

[20] Zhang ZC, Luo MK. New integral transforms for generalizing the Wigner distribution and ambiguity function. IEEE Signal Processing Letters. 2015;**22**: 460-464. DOI: 10.1109/ LSP.2014.2362616

[21] Zhang ZC. New Wigner distribution and ambiguity function based on the generalized translation in the linear canonical transform domain. Signal Processing. 2016;**118**:51-61. DOI: 10.1016/j.sigpro.2015.06.010

[22] Zhang ZC. Unified Wigner-Ville distribution and ambiguity function in the linear canonical transform domain. Signal Processing. 2015;**114**:45-60. DOI: 10.1016/j.sigpro.2015.02.016

[23] Zhang ZC. Novel Wigner distribution and ambiguity function associated with the linear canonical transform. Optik. 2016;**127**:4995-5012. DOI: 10.1016/j.ijleo.2016.02.028

[24] Fan XL, Kou KI, Liu MS. Quaternion Wigner-Ville distribution associated with the linear canonical transforms. Signal Processing. 2017;**130**:129-141. DOI: 10.1016/j.sigpro.2016.06.018

[25] Zhuo ZH, Zhong N, Xie YA, Xu Z. Entropic uncertainty relations of the Wigner-Ville distribution in linear canonical transform domain. Transactions of Beijing Institute of Technology. 2017;**37**:948-952. DOI: 10.15918/j. tbit1001-0645.2017.09.012

[26] Liu L, Luo M, Lai L. Instantaneous frequency estimation based on the Wigner-Ville distribution associated with linear canonical transform (WDL). Chinese Journal of Electronics. 2018;**27**: 123-127. DOI: 10.1049/cje.2017.07.009

[27] Xin HC, Li BZ. On a new Wigner-Ville distribution associated with linear canonical transform. EURASIP Journal on Advances in Signal Processing. 2021; **56**. DOI: 10.1186/s13634-021-00753-3

[28] Wigner EP. On the quantum correction for thermodynamic equilibrium. Physics Review. 1932;**40**: 749-759. DOI: 10.1103/PhysRev.40.749

[29] Cohen L. Time-frequency distributions–A review. Proceedings of the IEEE. 1989;**77**:941-981. DOI: 10.1109/ 5.30749

[30] Debnath L. Recent developments in the Wigner-Ville distribution and time frequency signal analysis. PINSA. 2002; **68A**:35-56. DOI: 10.1007/978-0-8176-8418-1_5

[31] Ville J. Theorie et applications de la notion de signal analytique. Cables et Transmissions. 1948;**2**:61-74

[32] Hennelly BM, Sheridan JT. Generalizing, optimizing, and inventing numerical algorithms for the fractional Fourier, Fresnel, and linear canonical transforms. Journal of the Optical Society of America. A. 2005;**22**:917-927. DOI: 10.1364/JOSAA.22.000917

[33] Nazarathy M, Shamir J. First-order optics: Operator representation for systems with loss or gain. Journal of the Optical Society of America. 1982;**72**: 1398-1408. DOI: 10.1364/JOSA.72.001398

[34] Bernardo LM. ABCD matrix formalism of fractional Fourier optics. Optical Engineering. 1996;**35**:732-740. DOI: 10.1117/1.600641

[35] Abe S, Sheridan JT. Optical operations on wave functions as the Abelian subgroups of the special affine Fourier transformation. Optics Letters. 1994;**19**:1801-1803. DOI: 10.1364/OL.19.001801

[36] Bhandari A, Zayed AI. Convolution and product theorem for the special affine Fourier transform. In: Nashed MZ, Li X, editors. Frontiers in Orthogonal Polynomials and q-Series. 2018. pp. 119-137. DOI: 10.1142/9789813228887_0006

[37] Zhi XY, Wei DY, Zhang W. A generalized convolution theorem for the special affine Fourier transform and its application to filtering. Optik. 2016;**127**: 2613-2616. DOI: 10.1016/j.ijleo.2015.11.211

[38] Pei SC, Ding JJ. Eigenfunctions of the offset Fourier, fractional Fourier, and linear canonical transforms. Journal of the Optical Society of America A. 2003;**20**:522-532. DOI: 10.1364/JOSAA.20.000522

[39] Bahri M, Muh. Saleh Arif F. Relation between quaternion Fourier transform and quaternion Wigner-Ville distribution associated with linear canonical transform. Journal of Applied Mathematics. 2017;**1-10**:1-10. DOI: 10.1155/2017/32473641

[40] El Kassimi M, El Haoui Y, Fahlaoui S. The Wigner-Ville distribution associated with the quaternion offset linear canonical transform. Analysis Mathematica. 2019; **45**:787-802. DOI: 10.1007/s10476-019-0007-0

[41] Bhat MY, Dar AH. Convolution and correlation theorems for Wigner-Ville distribution associated with the quaternion offset linear canonical transform. Signal, Image and Video Processing. 2022;**16**:1235-1242. DOI: 10.1007/s11760-021-02074-2

[42] Urynbassarova D, El Haoui Y, Zhang F. Uncertainty principles for Wigner-Ville distribution associated with the quaternion offset linear canonical transform. Circuits Systems and Signal Processing. 2022. DOI: 10.1007/s00034-022-02127-y

[43] Bhat MY, Dar AH. Wigner distribution and associated uncertainty principles in the framework of octonion linear canonical transform. DOI: 10.48550/arXiv.2209.05697

Chapter 3

Scaled Ambiguity Function Associated with Quadratic-Phase Fourier Transform

Mohammad Younus Bhat, Aamir Hamid Dar, Altaf Ahmad Bhat and Deepak Kumar Jain

Abstract

Quadratic-phase Fourier transform (QPFT) as a general integral transform has been considered into Wigner distribution (WD) and Ambiguity function (AF) to show more powerful ability for non-stationary signal processing. In this article, a new version of ambiguity function (AF) coined as scaled ambiguity function associated with the Quadratic-phase Fourier transform (QPFT) is proposed. This new version of AF is defined based on the QPFT and the fractional instantaneous autocorrelation. Firstly, we define the scaled ambiguity function associated with the QPFT (SAFQ). Then, the main properties including the conjugate-symmetry, shifting, scaling, marginal and Moyal's formulae of SAFQ are investigated in detail, the results show that SAFQ can be viewed as the generalization of the classical AF. Finally, the newly defined SAFQ is used for the detection of linear-frequency-modulated (LFM) signals.

Keywords: ambiguity function, quadratic-phase Fourier transform, Moyal's formula, modulation, linear frequency-modulated signal

1. Introduction

The Fourier transform is indeed an indispensable tool for the time-frequency analysis of the stationary signals. Due to its success stories FT has profoundly influenced the mathematical, biological, chemical and engineering communities over decades, but FT can not analyze non-stationary signals as it can not provide any valid information despite the localization properties of the spectral contents. FT only allows us to visualize the signals either in time or frequency domain, but not in both domains simultaneously. In Refs. [1–3], Castro et al. introduced a superlative generalized version of the Fourier transform(FT) called quadratic-phase Fourier transform(QPFT), which not only treats uniquely both the transient and non-transient signals in a nice fashion but also with non-orthogonal directions. The QPFT is actually a generalization of several well known transforms like Fourier, fractional Fourier and linear canonical transforms, offset linear canonical transform whose kernel is in the exponential form.

Many researches have been carried on quadratic-phase Fourier transform(see [4, 5]). With the fact that the QPFT is monitored by a bunch of free parameters, it has evolved as an effective tool for the representation of signals. A notable consideration has been given in the extension of the Wigner distributions to the classical QPFT and its generalizations. More can be found in Refs. [6–9].

On the other hand, the classical ambiguity function (AF) and Wigner distribution (WD) are the basic parametric time-frequency analysis tools, evolved for the analysis of time-frequency characteristics of non-stationary signals [10–14]. At the same time, the linear frequency-modulated (LFM) signal, a typical non-stationary signal, is widely used in communications, radar and sonar system. Many algorithms and methods have been proposed in view of LFM. The most important among them are the AF and WD [10, 13, 15–19], defined as the Fourier transform of the classical instantaneous autocorrelation function $\omega\left(t+\frac{\tau}{2}\right)\omega^*\left(t-\frac{\tau}{2}\right)$ for t and τ, (superscript $*$ denotes complex conjugate) respectively. It is well known that the AF offers perfect localization (localized on a straight line) to the mono-component LFM signals but cross terms appear while dealing with multi-component LFM signals as they are quadratic in nature. However these cross terms become troublesome if the frequency rate of one component approaches other. This drawback of AF gave rise to a series of different classes of time- frequency representation tools (see [20–27]). In Ref. [28], authors used fractional instantaneous auto-correlation $\omega\left(t+k\frac{\tau}{2}\right)\omega^*\left(t-k\frac{\tau}{2}\right)$ found in the definition of fractional bi-spectrum [29], which is parameterized by a constant $k\in\mathbb{Q}^+$ to introduced a scaled version of the conventional WD. Later Dar and Bhat [30] introduced the scaled version of Ambiguity function and Wigner distribution in the linear canonical transform domain. They also introduced scaled version of Wigner distribution in the offset linear canonical transform [31–35], hence provides a novel way for the improvement of the cross-term reduction time–frequency resolution and angle resolution.

Keeping in mind the degree of freedom corresponding to the choice of a factor k in the fractional instantaneous auto-correlation and the extra degree of freedom present in QPFT, we introduce a novel scaled ambiguity function in the quadratic-phase Fourier transform domain (SAFQ), which gives a unique treatment for all classical classes of AF's. Hence, it is good to study rigorously the SAFQ which will be effective for signal processing theory and applications especially for detection and estimation of LFM signals.

1.1 Paper contributions

The contributions of this paper are summarized below:

- To introduce a scaled ambiguity function associated with the quadratic-phase Fouricr transform.
- To study the fundamental properties of the SAFQ, including the conjugate symmetry, time marginal, non-linearity, time shift, frequency shift, frequency marginal, scaling and Moyal formula.
- To show the of advantage of the theory, we provide the applications of the proposed distribution in the detection of single-component and bi-component linear-frequency-modulated (LFM) signal.

1.2 Paper outlines

The paper is organized as follows: In Section 2, we gave a brief review of QPFT and introduce AF associated with it. The definition and the properties of the SAFQ are studied in Section 3. In Section 4, the applications of the proposed distribution for the detection of single-component and bi-component LMF signals is provided. Finally, a conclusion is drawn in Section 5.

2. Preliminary

In this section, we gave the definitions of the Quadratic-phase Fourier transform (QPFT), the ambiguity function associated with QPFT and the scaled ambiguity function which will be needed throughout the paper.

2.1 Quadratic-phase Fourier transform (QPFT)

For a given set of parameters of $\Omega = (A, B, C, D, E), B \neq 0$ the quadratic-phase Fourier transform any signal $\omega(t)$ is defined by [1–3]

$$\mathcal{Q}^{\Omega}[\omega](u) = \int_{\mathbb{R}} \omega(t) K_{\Omega}(t, u) dt, \tag{1}$$

where the quadratic-phase Fourier kernel $K_{\Omega}(t, w)$ is given by

$$K_{\Omega}(t, u) = \sqrt{\frac{B}{2\pi i}} e^{\left(At^2+Btu+Cu^2+Dt+Eu\right)}, \quad A,B,C,D.E \in \mathbb{R}. \tag{2}$$

2.2 Ambiguity function in the quadratic-phase fourier domain (AFQ)

Authors in Refs. [7, 8] defined the AF associated with the LCT, using the same procedure we can define the AF associated with QPFT (AFQ) as

$$AFQ^{\Omega}_{\omega(t)}(\tau, u) = \int_{\mathbb{R}} \omega\left(t + \frac{\tau}{2}\right) \omega^* \left(t - \frac{\tau}{2}\right) K_{\Omega}(\tau, u) dt, \tag{3}$$

2.3 Scaled ambiguity function

For a finite energy signal the scaled Ambiguity function (SAF) is defined as Ref. [30].

$$SAF_{\omega(t)}(\tau, u) = \int_{\mathbb{R}} \omega\left(t + k\frac{\tau}{2}\right) \omega^* \left(t - \frac{\tau}{2}\right) e^{-iut} dt, \tag{4}$$

where $k \in \mathbb{Q}^+$ the set of positive rational numbers.

3. Scaled ambiguity function associated with quadratic-phase fourier transform (SAFQ)

In this section, we shall introduce the notion of the scaled Ambiguity function associated with QPFT followed by some of its basic properties.

3.1 Definition of the scaled AFQ

Thanks to the scaled AF, we obtain obtain different expressions for the SAFQ as follows:

$$\begin{aligned} SAF_{\omega(t)}(\tau, u) &= \int_{\mathbb{R}} \omega\left(t + k\frac{\tau}{2}\right)\omega^*\left(t - k\frac{\tau}{2}\right)e^{-iut}dt \\ &= \int_{\mathbb{R}} \omega\left(t + k\frac{\tau}{2}\right)e^{-i\frac{u}{2}\left(t+k\frac{\tau}{2}\right)}\omega^*\left(t - \frac{\tau}{2}\right)e^{-i\frac{u}{2}\left(t-k\frac{\tau}{2}\right)}dt \\ &= \int_{\mathbb{R}} \overline{\omega}_u\left(t + k\frac{\tau}{2}\right)\hat{\omega}_u^*\left(t - k\frac{\tau}{2}\right)dt, \end{aligned} \tag{5}$$

where

$$\overline{\omega}_u(t) = \omega(t)e^{-i\frac{u}{2}t} \quad \text{and} \quad \hat{\omega}_u(t) = \omega(t)e^{i\frac{u}{2}t}. \tag{6}$$

On replacing the Fourier kernel in (6) with the QPFT kernel, we obtain

$$\overline{\omega}_u^{\Omega}(t) = \omega(t)K_{\Omega}\left(t, \frac{u}{2}\right) \quad \text{and} \quad \hat{\omega}_u^{\Omega}(t) = \omega(t)K_{\Omega}\left(t, -\frac{u}{2}\right). \tag{7}$$

Thus, we obtain a new version of scaled AF associated with the QPFT by replacing $\overline{\omega}_u(t)$ with $\overline{\omega}_u^{\Omega}(t)$ and $\hat{\omega}_u(t)$ with $\hat{\omega}_u^{\Omega}(t)$ in (5), i.e.,

$$\begin{aligned} SAF^{\Omega}_{\omega(t)}(\tau, u) &= \int_{\mathbb{R}} \overline{\omega}_u^{\Omega}\left(t + k\frac{\tau}{2}\right)\hat{x}_u^{\Omega\,*}\left(t - k\frac{\tau}{2}\right)dt \\ &= \int_{\mathbb{R}} \omega\left(t + k\frac{\tau}{2}\right)K_{\Omega}\left(t + k\frac{\tau}{2}, \frac{u}{2}\right)\omega^*\left(t - k\frac{\tau}{2}\right)K_{\Omega}^*\left(t - k\frac{\tau}{2}, \frac{-u}{2}\right)dt \\ &= \frac{B}{2\pi}\int_{\mathbb{R}} \omega\left(t + k\frac{\tau}{2}\right)\omega^*\left(t - k\frac{\tau}{2}\right)e^{i[(2Ak\tau+Bu)t+Dk\tau+Eu]}dt. \end{aligned} \tag{8}$$

With the virtue of above equation we have following definition.

Definition 3.1. *The scaled Ambiguity function associated with quadratic-phase Fourier transform of a signal '$\omega(t)$' in $L^2(\mathbb{R})$ with respect the real parameter set $\Omega = (A, B, C, D, E)$,$B \neq 0$ is defined as*

$$SAF^{\Omega}_{\omega(t)}(\tau, u) = \frac{B}{2\pi}\int_{\mathbb{R}} \omega\left(t + k\frac{\tau}{2}\right)\omega^*\left(t - k\frac{\tau}{2}\right)e^{i[(2Ak\tau+Bu)t+Dk\tau+Eu]}dt, \tag{9}$$

where $k \in \mathbb{Q}^+$.

It is worth to mention that if we change the parameter $\Omega = (A, B, C, D, E)$ in the Definition 3.1, we have the following important deductions:

i. When the parameter $\Omega = (A/2B, -1/B, C/2B, 0, 0)$ is chosen and multiplying the right side of (9) by -1, the SAFQ (9) yields the scaled ambiguity function associated with linear canonical transform [30]:

$$SAF^{\Omega}_{\omega(t)}(\tau, u) = \frac{1}{2\pi B}\int_{\mathbb{R}} \omega\left(t + k\frac{\tau}{2}\right)\omega^*\left(t - k\frac{\tau}{2}\right)e^{i\frac{1}{B}(Ak\tau - u)t}dt. \tag{10}$$

ii. For the set $\Omega = (\cot\zeta/2, -\csc\zeta, \cot\zeta/2, 0, 0), \zeta \neq 2\pi$ and multiplying the right side of (9) by -1 the SAFQ (9) yields the novel scaled AF associated with fractional Fourier transform:

$$SAF^{\zeta}_{\omega(t)}(t, u) = \frac{1}{2\pi \sin\zeta}\int_{\mathbb{R}} \omega\left(t + k\frac{\theta}{2}\right)\omega^*\left(t - k\frac{\tau}{2}\right)e^{i((k\cot\zeta\tau - u\csc\zeta)t}dt. \tag{11}$$

iii. When the parameter is choosen as $\Omega = (0,1,0,0,0)$ is chosen, the scaled AFQ (4) boils down to the classical scaled AF given in Ref. [30]. In addition of above if we take $k = 1$, it reduce to classical Amniguity function.

3.2 Properties of the scaled AFOL

In this subsection, we investigate some general properties of the scaled AFQ with their detailed proofs. These properties play vital role in signal representation. We shall see the differences between the scaled versions and conventional ones.

Property 3.1 (symmetry property) *For $\omega(t) \in L^2(\mathbb{R})$, then scaled AFOL of the signals $\omega^*(t)$ and $P[\omega(t)]$ have the following forms*

$$SAF^{\Omega}_{\omega(t)^*}(\tau, u) = SAF^{\Omega'}_{\omega(t)}(-\tau, -u) \tag{12}$$

where $\Omega' = (-A. - B, C, -D, -E)$.
and

$$SAF^{\Omega}_{P[\omega(t)]}(\tau, u) = -SAF^{\overline{\Omega}}_{\omega(t)}(-\tau, -u), \tag{13}$$

where $P[\omega(t)] = \omega(-t)$ and $\overline{\Omega} = (A, B, C, -D, -E)$.
Proof. From Definition 3.1, we have

$$SAF^{\Omega}_{\omega(t)^*}(\tau, u)$$

$$= \frac{B}{2\pi}\int_{\mathbb{R}} \omega^*\left(t + k\frac{\tau}{2}\right)\omega\left(t - k\frac{\tau}{2}\right)e^{i[(2Ak\tau + Bu)t + Dk\tau + Eu]}dt$$

$$= \frac{B}{2\pi}\int_{\mathbb{R}} \omega\left(t + k\frac{(-\tau)}{2}\right)\omega^*\left(t - k\frac{(-\tau)}{2}\right)e^{i[(2Ak\tau + Bu)t + Dk\tau + Eu]}dt$$

$$= \frac{B}{2\pi}\int_{\mathbb{R}} \omega\left(t + k\frac{(-\tau)}{2}\right)\omega^*\left(t - k\frac{(-\tau)}{2}\right) \times e^{i[\{2(-A)k(-\tau) + (-B)(-u)\}t + (-D)k(-\tau) + (-E)(-u)]}dt$$

$$= SAF^{\Omega'}_{\omega(t)}(-\tau, -u), \quad \textit{where} \quad \Omega' = (-A. - B, C, -D, -E).$$

which prove (12).
Now, we move forward to prove (13)
From (9), we have

$$
\begin{aligned}
&SAF^{\Omega}_{P[\omega(t)]}(\tau, u)\\
&= \frac{B}{2\pi}\int_{\mathbb{R}} P\omega\left(t+k\frac{\tau}{2}\right)P\omega^*\left(t-k\frac{\tau}{2}\right)e^{i[(2Ak\tau+Bu)t+Dk\tau+Eu]}dt\\
&= \frac{B}{2\pi}\int_{\mathbb{R}} \omega\left(-t-k\frac{\tau}{2}\right)\omega^*\left(-t+k\frac{\tau}{2}\right)e^{i[(2Ak\tau+Bu)t+Dk\tau+Eu]}dt\\
&= \frac{B}{2\pi}\int_{\mathbb{R}} \omega\left(-t-k\frac{\tau}{2}\right)\omega^*\left(-t+k\frac{\tau}{2}\right)e^{i[\{2Ak(-\tau)+B(-u)\}(-t)+(-D)k(-\tau)+(-E)(-u)]}dt\\
&= -\frac{B}{2\pi}\int_{\mathbb{R}} \omega\left(v+k\frac{-\tau}{2}\right)\omega^*\left(v-k\frac{-\tau}{2}\right)e^{i[\{2Ak(-\tau)+B(-u)\}v+(-D)k(-\tau)+(-E)(-u)]}dv\\
&= -SAF^{\overline{\Omega}}_{\omega(t)}(-\tau, -u), \quad \overline{\Omega} = (A, B, C, -D, -E).
\end{aligned}
$$

which completes the proof. □

Property 3.2 (Time shift). *The SAFQ of a signal $\omega(t-\lambda)$ can be expressed as:*

$$
SAF^{\Omega}_{\omega(t-\lambda)}(\tau, u) = e^{i\lambda(2Ak\tau+Bu)}SAF^{\Omega}_{\omega(t)}(\tau, u). \tag{14}
$$

Proof. From (9), we obtain

$$
SAF^{\Omega}_{\omega(t-\lambda)}(\tau, u) = \frac{B}{2\pi}\int_{\mathbb{R}} \omega\left(t-\lambda+k\frac{\tau}{2}\right)\omega^*\left(t-\lambda-k\frac{\tau}{2}\right)e^{i[(2Ak\tau+Bu)t+Dk\tau+Eu]}dt.
$$

Setting $t-\lambda = s$, we have from last equation

$$
\begin{aligned}
SAF^{\Omega}_{\omega(t-\lambda)}(\tau, u) &= \frac{B}{2\pi}\int_{\mathbb{R}} \omega\left(s+k\frac{\tau}{2}\right)\omega^*\left(s-k\frac{\tau}{2}\right)e^{i[(2Ak\tau+Bu)s+Dk\tau+Eu]}ds\\
&= e^{i\lambda(2Ak\tau+Bu)}\frac{B}{2\pi}\int_{\mathbb{R}} \omega\left(s+k\frac{\tau}{2}\right)\omega^*\left(s-k\frac{\tau}{2}\right)e^{\frac{i}{b}[(ak\tau-u)s+ku_0\tau-u(du_0-bw_0)]}ds\\
&= e^{i\lambda(2Ak\tau+Bu)}SAF^{\Omega}_{\omega(t)}(\tau, u).
\end{aligned}
$$

Which completes the proof of (14). □

Property 3.3 (Frequency shift). *The SAFQ of a signal $\omega(t)e^{ivt}$ can be expressed as:*

$$
SAF^{\Omega}_{\omega(t)e^{ivt}}(\tau, u) = e^{ivk\tau}SAF^{\Omega}_{\omega(t)}(\tau, u) \tag{15}
$$

Proof. From (9), we have

$$
\begin{aligned}
SAF^{\Omega}_{\omega(t)e^{ivt}}(\tau, u) &= \frac{B}{2\pi}\int_{\mathbb{R}} \omega\left(t+k\frac{\tau}{2}\right)e^{iv\left(t+k\frac{\tau}{2}\right)}\omega^*\left(t-\frac{\tau}{2}\right)e^{-iv\left(t-k\frac{\tau}{2}\right)}\\
&\qquad\times e^{i[(2Ak\tau+Bu)t+Dk\tau+Eu]}dt\\
&= \frac{B}{2\pi}\int_{\mathbb{R}} \omega\left(t+k\frac{\tau}{2}\right)\omega^*\left(t-\frac{\tau}{2}\right)e^{ivk\tau}\\
&\qquad\times e^{i[(2Ak\tau+Bu)t+Dk\tau+Eu]}dt\\
&= e^{ivk\tau}\frac{B}{2\pi}\int_{\mathbb{R}} \omega\left(t+k\frac{\tau}{2}\right)\omega^*\left(t-\frac{\tau}{2}\right)\\
&\qquad\times e^{i[(2Ak\tau+Bu)t+Dk\tau+Eu]}dt\\
&= e^{ivk\tau}SAF^{\Omega}_{\omega(t)}(\tau, u).
\end{aligned}
$$

Which completes the proof □

Property 3.4 (Non-linearity). *Let $\omega(t)=\omega_1(t)+\omega_2(t)$ be in $L^2(\mathbb{R})$, then we have*

$$SAF^{\Omega}_{\omega(t)}(\tau,u)=SAF^{\Omega}_{\omega_1(t)}(\tau,u)+SAF^{\Omega}_{\omega_2(t)}(\tau,u)+SAF^{\Omega}_{\omega_1,\omega_2}(\tau,u)+SAF^{\Omega}_{\omega_2,\omega_1}(\tau,u) \tag{16}$$

Proof. From Definition 3.1, we have

$$\begin{aligned}
&SAF^{\Omega}_{\omega(t)}(\tau,u)\\
&=\frac{B}{2\pi}\int_{\mathbb{R}}(\omega_1+\omega_2)\left(t+k\frac{\tau}{2}\right)(\omega_1+\omega_2)^*\left(t-k\frac{\tau}{2}\right)e^{i[(2Ak\tau+Bu)t+Dk\tau+Eu]}dt\\
&=\frac{B}{2\pi}\int_{\mathbb{R}}\left[\left(\omega_1\left(t+k\frac{\tau}{2}\right)+\omega_2\left(t+k\frac{\tau}{2}\right)\right)\right.\\
&\qquad\left.\left(\omega_1{}^*\left(t-k\frac{\tau}{2}\right)+\omega_2{}^*\left(t-k\frac{\tau}{2}\right)\right)\right]e^{i[(2Ak\tau+Bu)t+Dk\tau+Eu]}dt\\
&=\frac{B}{2\pi}\int_{\mathbb{R}}\left[\omega_1\left(t+k\frac{\tau}{2}\right)\omega_1{}^*\left(t-k\frac{\tau}{2}\right)+\omega_2\left(t+k\frac{\tau}{2}\right)\omega_2{}^*\left(t-k\frac{\tau}{2}\right)\right.\\
&\qquad\left.+\omega_1\left(t+k\frac{\tau}{2}\right)\omega_2{}^*\left(t-k\frac{\tau}{2}\right)+\omega_2\left(t+k\frac{\tau}{2}\right)\omega_1{}^*\left(t-k\frac{\tau}{2}\right)\right]\\
&\qquad\qquad\times e^{i[(2Ak\tau+Bu)t+Dk\tau+Eu]}dt\\
&=SAF^{\Omega}_{\omega_1}(\tau,u)+SAF^{\Omega}_{\omega_2}(\tau,u)+SAF^{\Omega}_{\omega_1,\omega_2}(\tau,u)+SAF^{\Omega}_{\omega_2,\omega_1}(\tau,u).
\end{aligned}$$

Thus completes the proof. □

Property 3.5 (Frequency marginal property). *The frequency marginal property of SAFQ is given by*

$$\int_{\mathbb{R}}SAF^{\Omega}_{\omega(t)}(\tau,u)d\tau=\frac{1}{k}\mathcal{Q}^{\Omega}[\omega(t)]\left(\frac{u}{2}\right)\mathcal{Q}^{*\Omega}[\omega(t)]\left(\frac{-u}{2}\right) \tag{17}$$

Proof. From Definition 3.1, we have

$$\int_{\mathbb{R}}SAF^{\Omega}_{\omega(t)}(\tau,u)d\tau=\frac{B}{2\pi}\int_{\mathbb{R}^2}\omega\left(t+k\frac{\tau}{2}\right)\omega^*\left(t-\frac{\tau}{2}\right)e^{i[(2Ak\tau+Bu)t+Dk\tau+Eu]}dtd\tau.$$

Making change of variable $t+k\frac{\tau}{2}=s$, above equation yields

$$\int_{\mathbb{R}}SAF^{\Omega}_{\omega(t)}(\tau,u)d\tau=\frac{B}{\pi k}\int_{\mathbb{R}^2}\omega(s)\omega^*(2t-s)e^{i[\{4A(s-t)+Bu\}t+2D(s-t)+Eu]}dsdt.$$

Now setting $2t=s+v$, we get

$$\begin{aligned}
&\int_{\mathbb{R}} SAF^{\Omega}_{\omega(t)}(\tau, u)d\tau \\
&= \frac{B}{2\pi k}\int_{\mathbb{R}^2} \omega(s)\omega^*(v)e^{i\left[\left\{4A\left(s-\frac{s+v}{2}\right)-Bu\right\}\left(\frac{s+v}{2}\right)+2D\left(s-\frac{s+v}{2}\right)+Eu\right]}dsdv \\
&= \frac{B}{2\pi k}\int_{\mathbb{R}^2} \omega(s)\omega^*(v)e^{i\left[\{2A(s-v)+Bu\}\left(\frac{s+v}{2}\right)+D(s-v)+Eu\right]}dsdv \\
&= \frac{B}{2\pi k}\int_{\mathbb{R}^2} \omega(s)\omega^*(v)e^{i\left[A\left(s^2-v^2\right)+Bu\left(\frac{s+v}{2}\right)+D(s-v)+Eu\right]}dsdv \\
&= \frac{B}{2\pi k}\int_{\mathbb{R}} \omega(s)e^{i\left[As^2+Bs\left(\frac{u}{2}\right)+C\left(\frac{u}{2}\right)^2+Ds+E\left(\frac{u}{2}\right)\right]}ds \\
&\qquad \times\int_{\mathbb{R}} \omega^*(v)e^{-i\left[Av^2+Bv\left(\frac{-u}{2}\right)+C\left(\frac{-u}{2}\right)^2+Dv+E\left(\frac{-u}{2}\right)\right]}dv \\
&= \frac{1}{k}\int_{\mathbb{R}} \omega(s)\sqrt{\frac{B}{2i\pi}}e^{i\left[As^2+Bs\left(\frac{u}{2}\right)+C\left(\frac{u}{2}\right)^2+Ds+E\left(\frac{u}{2}\right)\right]}ds \\
&\qquad \times\left[\int_{\mathbb{R}} \omega(v)\sqrt{\frac{B}{2i\pi}}e^{i\left[Av^2+Bv\left(\frac{-u}{2}\right)+C\left(\frac{-u}{2}\right)^2+Dv+E\left(\frac{-u}{2}\right)\right]}dv\right]^* \\
&= \frac{1}{k}\int_{\mathbb{R}} \omega(s)K_{\Omega}\left(s, \frac{u}{2}\right)ds\left[\int_{\mathbb{R}} \omega(v)K_{\Omega}\left(v, \frac{-u}{2}\right)dv\right]^* \\
&= \frac{1}{k}\mathcal{Q}^{\Omega}[\omega(t)]\left(\frac{u}{2}\right)\mathcal{Q}^{*\Omega}[\omega(t)]\left(\frac{-u}{2}\right).
\end{aligned}$$

Which completes the proof. □

Property 3.6 (Scaling property). *For a signal $\tilde{\omega}(t) = \sqrt{\sigma}\omega(\sigma t)$ the SAFQ has the following form:*

$$SAF^{\Omega}_{\tilde{\omega}(t)}(\tau, u) = SAF^{\Omega'}_{\omega(t)}\left(\sigma\tau, \frac{u}{\sigma}\right), \tag{18}$$

where $\Omega' = \left(\frac{A}{\sigma^2}, B, C, \frac{D}{\sigma}, \sigma E\right)$.

Proof. From (9), we have

$$SAF^{\Omega}_{\tilde{\omega}(t)}(\tau, u) = \frac{\sigma B}{2\pi}\int_{\mathbb{R}} \omega\left(\sigma t + \sigma k\frac{\tau}{2}\right)\omega^*\left(\sigma t - \sigma k\frac{\tau}{2}\right)e^{i[(2Ak\tau+Bu)t+Dk\tau+Eu]}dt.$$

Setting $\sigma t = \eta$, above equation yields

$$\begin{aligned}
&SAF^{\Omega}_{\tilde{\omega}(t)}(\tau, u) \\
&= \frac{\sigma B}{2\pi}\int_{\mathbb{R}} \omega\left(\sigma t + \sigma k\frac{\tau}{2}\right)\omega^*\left(\sigma t - \sigma k\frac{\tau}{2}\right)e^{i\left[(2Ak\tau+Bu)\frac{\eta}{\sigma}+Dk\tau+Eu\right]}.\frac{d\eta}{\sigma} \\
&= \frac{\sigma B}{2\pi}\int_{\mathbb{R}} \omega\left(\sigma t + \sigma k\frac{\tau}{2}\right)\omega^*\left(\sigma t - \sigma k\frac{\tau}{2}\right)e^{i\left[\left(2\frac{A}{\sigma^2}k(\sigma\tau)+B\frac{u}{\sigma}\right)\eta+Dk\tau+Eu\right]}.\frac{d\eta}{\sigma} \\
&= \frac{B}{2\pi}\int_{\mathbb{R}} \omega\left(\sigma t + \sigma k\frac{\tau}{2}\right)\omega^*\left(\sigma t - \sigma k\frac{\tau}{2}\right)e^{i\left[\left(2\frac{A}{\sigma^2}k(\sigma\tau)+B\left(\frac{u}{\sigma}\right)\right)\eta+\frac{D}{\sigma}k(\sigma\tau)+\sigma E\left(\frac{u}{\sigma}\right)\right]}.d\eta \\
&= SAF^{\Omega'}_{\omega(t)}\left(\sigma\tau, \frac{u}{\sigma}\right),
\end{aligned}$$

where $\Omega' = \left(\frac{A}{\sigma^2}, B, C, \frac{D}{\sigma}, \sigma E\right)$.
This proves (18). □

Property 3.7 (Moyal formula). *The Moyal formula of the SAFQ has the following form:*

$$\int_{\mathbb{R}}\int_{\mathbb{R}} SAF^{\Omega}_{\omega_1(t)}(\tau, u)\left[SAF^{\Omega}_{\omega_2(t)}(\tau,\ u)\right]^* d\tau du = \frac{B}{2\pi k}|\langle\omega_1(t), \omega_2(t)\rangle|^2. \tag{19}$$

Proof. From (9), we have

$$\begin{aligned}
&\int_{\mathbb{R}}\int_{\mathbb{R}} SAF^{\Omega}_{\omega_1(t)}(t, u)\left[SAF^{\Omega}_{\omega_2(t)}(t,\ u)\right]^* dtdu \\
&= \left(\frac{B}{2\pi}\right)^2 \int_{\mathbb{R}}\int_{\mathbb{R}}\int_{\mathbb{R}}\int_{\mathbb{R}} \omega_1\left(t+k\frac{\tau}{2}\right)\omega_1^*\left(t-k\frac{\tau}{2}\right)\omega_2^*\left(t'+k\frac{\tau}{2}\right)\omega_2\left(t'-k\frac{\tau}{2}\right) \\
&\qquad \times e^{i[(2Ak\tau+Bu)t+Dk\tau+Eu]}e^{-i[(2Ak\tau+Bu)t'+Dk\tau+Eu]}d\tau dt'dtdu \\
&= \left(\frac{B}{2\pi}\right)^2 \int_{\mathbb{R}}\int_{\mathbb{R}}\int_{\mathbb{R}}\int_{\mathbb{R}} \omega_1\left(t+k\frac{\tau}{2}\right)\omega_1^*\left(t-\frac{\tau}{2}\right)\omega_2^*\left(t+k\frac{\tau}{2}\right)\omega_2\left(t-k\frac{\tau}{2}\right) \\
&\qquad \times e^{i(2Ak\tau+Bu)(t-t')}d\tau dudtdt' \\
&= \frac{B}{2\pi}\int_{\mathbb{R}}\int_{\mathbb{R}}\int_{\mathbb{R}} \omega_1\left(t+k\frac{\tau}{2}\right)\omega_1^*\left(t-\frac{\tau}{2}\right)\omega_2^*\left(t+k\frac{\tau}{2}\right)\omega_2\left(t-k\frac{\tau}{2}\right) \\
&\qquad \times e^{i2Ak\tau(t-t')}\left(\frac{B}{2\pi}\int_{\mathbb{R}} e^{iBu(t-t')}du\right)d\tau dtdt' \\
&= \frac{B}{2\pi}\int_{\mathbb{R}}\int_{\mathbb{R}}\int_{\mathbb{R}} \omega_1\left(t+k\frac{\tau}{2}\right)\omega_1^*\left(t-\frac{\tau}{2}\right)\omega_2^*\left(t+k\frac{\tau}{2}\right)\omega_2\left(t-k\frac{\tau}{2}\right) \\
&\qquad \times e^{i2Ak\tau(t-t')}\delta(t-t')dt'd\tau dt \\
&= \frac{B}{2\pi}\int_{\mathbb{R}}\int_{\mathbb{R}} \omega_1\left(t+k\frac{\tau}{2}\right)\omega_1^*\left(t-\frac{\tau}{2}\right)\omega_2^*\left(t+k\frac{\tau}{2}\right)\omega_2\left(t-k\frac{\tau}{2}\right)d\tau dt
\end{aligned}$$

By making the change of variable $s = t + k\frac{\tau}{2}$, we have

$$\int_{\mathbb{R}}\int_{\mathbb{R}} \mathcal{W}^{A,k}_{\omega_1(t)}(t, u)\left[\mathcal{W}^{A,k}_{\omega_2(t)}(t,\ u)\right]^* d\tau du = \frac{B}{k\pi}\int_{\mathbb{R}}\int_{\mathbb{R}} \omega_1(s)\omega_1^*(2t-s)\omega_2^*(s)\omega_2(2t-s)dsdt$$

Now taking $2t - s = v$, we obtain

$$\begin{aligned}
\int_{\mathbb{R}}\int_{\mathbb{R}} \mathcal{W}^{A,k}_{\omega_1}(t, u)\left[\mathcal{W}^{A,k}_{\omega_2}(t,\ u)\right]^* d\tau du &= \frac{B}{2\pi k}\int_{\mathbb{R}}\int_{\mathbb{R}} \omega_1(s)\omega_1^*(v)\omega_2^*(s)\omega_2(v)dsdv \\
&= \frac{B}{2\pi k}\left(\int_{\mathbb{R}} \omega_1(s)\omega_2^*(s)dx\right)\left(\int_{\mathbb{R}} \omega_1^*(v)\omega_2(v)dv\right) \\
&= \frac{B}{2\pi k}|\langle\omega_1(t), \omega_2(t)\rangle|^2.
\end{aligned}$$

Thus completes the proof. □

4. Applications of the scaled AFQ

In engineering the most important research topics is the detection of LFM signals as they are widely used in communications, information and optical systems. In this section our main goal is to use scaled AFQ in detection of one-component and bi-component LFM signals, respectively.

- **One component LFM signal:** A one-component LFM signal is chosen as

$$\omega(t) = e^{i\left(\vartheta_1 t+\vartheta_2 t^2\right)} \tag{20}$$

where ϑ_1 and ϑ_2 represent the initial frequency and frequency rate of $\omega(t)$, respectively. Then, we obtain the SAFQ of a signal $\omega(t)$ as shown in the following theorem.

Theorem 4.1 *The SAFQ of* $\omega(t) = e^{i\left(\vartheta_1 t+\vartheta_2 t^2\right)}$ *can be presented as*

$$SAF^{\Omega}_{\omega(t)}(\tau, u) = e^{i[k(\vartheta_1+D)\tau+Eu]}\delta[2k(\vartheta_2 + A)\tau + Bu]. \tag{21}$$

Proof. By Definition 3.1, we have

$$\begin{aligned}
&SAF^{\Omega}_{\omega(t)}(\tau, u)\\
&= \frac{B}{2\pi}\int_{\mathbb{R}} \omega\left(t + k\frac{\tau}{2}\right)\omega^*\left(t - k\frac{\tau}{2}\right)e^{i[(2Ak\tau+Bu)t+Dk\tau+Eu]}dt\\
&= \frac{B}{2\pi}\int_{\mathbb{R}} e^{i\left[\vartheta_1\left(t+k\frac{\tau}{2}\right)+\vartheta_2\left(t+k\frac{\tau}{2}\right)^2\right]}e^{-i\left[\vartheta_1\left(t-k\frac{\tau}{2}\right)+\vartheta_2\left(t-k\frac{\tau}{2}\right)^2\right]}\\
&\qquad\times e^{i[(2Ak\tau+Bu)t+Dk\tau+Eu]}dt\\
&= \frac{B}{2\pi}\int_{\mathbb{R}} e^{i\left[\vartheta_1 t+\vartheta_1 k\frac{\tau}{2}+\vartheta_2 t^2+\vartheta_2 tk\tau+\vartheta_2 k^2\frac{\tau^2}{4}\right]}e^{-i\left[\vartheta_1 t-\vartheta_1 k\frac{\tau}{2}+\vartheta_2 t^2-\vartheta_2 tk\tau+\vartheta_2 k^2\frac{\tau^2}{4}\right]}\\
&\qquad\times e^{i[(2Ak\tau+Bu)t+Dk\tau+Eu]}dt\\
&= \frac{B}{2\pi}\int_{\mathbb{R}} e^{i[\vartheta_1 k\tau+2\vartheta_2 tk\tau+(2Ak\tau+Bu)t+Dk\tau+Eu]}dt\\
&= \frac{B}{2\pi}e^{i[k(\vartheta_1+D)\tau+Eu]}\int_{\mathbb{R}} e^{i[2k(\vartheta_2+A)\tau+Bu]t}dt\\
&= e^{i[k(\vartheta_1+D)\tau+Eu]}\delta[2k(\vartheta_2 + A)\tau + Bu],
\end{aligned} \tag{22}$$

□

From above Theorem, we can conclude that the that the SAFQ of a one-component signal (20) are able to generate impulses in (τ, u) plane at a straight line $(Bu + 2k(\vartheta_2 + A)\tau) = 0$ and is dependent on the scaling factor k and the parameter $\Omega = (A, B, C, D, E)$. Therefore, the SAFQ can be applied to the detection of one-component LFM signals and is very useful and effective as there is choice of selecting the scaling factor k and the parameter Ω.

- **Bi-component LFM signal:** Consider the following bi-component LFM signal $\omega(t)$ it is well known that the bi-component LFM signal can be expressed by the summation of two single component LFM signals, i.e.,

$$\omega(t) = \omega_1(t) + \omega_2(t), \tag{23}$$

where $\omega_1(t) = e^{i\left(\xi_1 t+\eta_1 t^2\right)}(\eta_1 \neq 0)$, $\omega_2(t) = e^{i\left(\xi_2 t+\eta_2 t^2\right)}(\eta_2 \neq 0)$ and $\eta_1 \neq \eta_2$. Now using the non-linearity property (8), the SAFQ of the signal $\omega(t)$ given in (23) can be computed as follows:

$$
\begin{aligned}
SAF^{\Omega}_{\omega(t)}(\tau, u) &= SAF^{\Omega}_{\omega_1(t)+\omega_2(t)}(\tau, u) \\
&= SAF^{\Omega}_{\omega_1(t)}(\tau, u) + SAF^{\Omega}_{\omega_2(t)}(\tau, u) + SAF^{\Omega}_{\omega_1(t),\omega_2(t)}(\tau, u) + SAF^{\Omega}_{\omega_2(t),\omega_1(t)}(\tau, u) \\
&= e^{i[k(\xi_1+D)\tau+Eu]}\delta[2k(\eta_1 + A)\tau + Bu] \\
&+e^{i[k(\xi_1+D)\tau+Eu]}\delta[2k(\eta_2 + A)\tau + Bu] + SAF^{\Omega}_{\omega_1(t),\omega_2(t)}(\tau, u) + SAF^{\Omega}_{\omega_2(t),\omega_1(t)}(\tau, u).
\end{aligned}
$$

The first two terms in last equation stands for the auto-terms of one-component signals, whereas the rest represent the cross terms that are given by

$$
\begin{aligned}
&SAF^{\Omega}_{\omega_1(t),\omega_2(t)}(t, u) \\
&= \frac{B}{2\pi}\int_{\mathbb{R}} \omega_1\left(t + k\frac{\tau}{2}\right)\omega_2^*\left(t - k\frac{\tau}{2}\right)e^{i[(2Ak\tau+Bu)t+Dk\tau+Eu]}dt \\
&= \frac{B}{2\pi}\int_{\mathbb{R}} e^{i\left[\xi_1\left(t+k\frac{\tau}{2}\right)+\eta_1\left(t+k\frac{\tau}{2}\right)^2\right]}e^{-i\left[\xi_2\left(t-k\frac{\tau}{2}\right)+\eta_2\left(t-k\frac{\tau}{2}\right)^2\right]}e^{i[(2Ak\tau+Bu)t+Dk\tau+Eu]}dt \\
&= \frac{B}{2\pi}\int_{\mathbb{R}} e^{i\left[\xi_1 t+\xi_1 k\frac{\tau}{2}+\eta_1 t^2+\eta_1 k^2\frac{\tau^2}{4}+\eta_1 tk\tau\right]}e^{-i\left[\xi_2 t-\xi_2 k\frac{\tau}{2}+\eta_2 t^2+\eta_2 k^2\frac{\tau^2}{4}-\eta_2 tk\tau\right]} \\
&\qquad\qquad \times e^{i[(2Ak\tau+Bu)t+Dk\tau+Eu]}dt \\
&= \frac{B}{2\pi}e^{i\left[\frac{\eta_1-\eta_2}{4}k^2\tau^2+\frac{\xi_1+\xi_2+2D}{2}k\tau+Eu\right]}\int_{\mathbb{R}} e^{i(\eta_1-\eta_2)t^2}e^{i[Bu+k(\eta_1+\eta_2+2A)\tau+(\xi_1-\xi_2)]t}dt \\
&= \frac{B}{k}\frac{1}{\sqrt{\pi(\eta_1-\eta_2)}}e^{i\left[\frac{\eta_1-\eta_2}{4}k^2\tau^2+\frac{\xi_1+\xi_2+2D}{2}k\tau+Eu\right]}e^{-i\frac{[Bu+k(\eta_1+\eta_2+2A)t-(\xi_1-\xi_2)]^2}{4(\eta_1-\eta_2)}},
\end{aligned}
$$

similarly

$$
\begin{aligned}
&SAF^{\Omega}_{\omega_2(t),\omega_1(t)}(t, u) \\
&= \frac{1}{kb}\frac{1}{\sqrt{\pi(\eta_2-\eta_1)}}e^{i\left[\frac{\eta_2-\eta_1}{4}k^2\tau^2+\frac{\xi_1+\xi_2+2d}{2}k\tau-Eu\right]}e^{-i\frac{[Bu+k(\eta_2+\eta_1+2A)t-(\xi_2-\xi_1)]^2}{4(\eta_2-\eta_1)}}.
\end{aligned}
$$

Hence the SAFQ of a bi-component signal $\omega(t) = \omega_1(t) + \omega_2(t)$ is given by

$$
\begin{aligned}
SAF^{\Omega}_{\omega(t)}(\tau, u) &= SAF^{\Omega}_{\omega_1(t)+\omega_2(t)}(\tau, u) \\
&= e^{i[k(\xi_1+D)\tau+Eu]}\delta[2k(\eta_1 + A)\tau + Bu] \\
&+e^{i[k(\xi_1+D)\tau+Eu]}\delta[2k(\eta_2 + A)\tau + Bu] \\
&+\frac{B}{k}\frac{1}{\sqrt{\pi(\eta_1-\eta_2)}}e^{i\left[\frac{\eta_1-\eta_2}{4}k^2\tau^2+\frac{\xi_1+\xi_2+2D}{2}k\tau+Eu\right]}e^{-i\frac{[Bu+k(\eta_1+\eta_2+2A)t-(\xi_1-\xi_2)]^2}{4(\eta_1-\eta_2)}} \\
&+\frac{1}{kb}\frac{1}{\sqrt{\pi(\eta_2-\eta_1)}}e^{i\left[\frac{\eta_2-\eta_1}{4}k^2\tau^2+\frac{\xi_1+\xi_2+2d}{2}k\tau-Eu\right]}e^{-i\frac{[Bu+k(\eta_2+\eta_1+2A)t-(\xi_2-\xi_1)]^2}{4(\eta_2-\eta_1)}}.
\end{aligned} \tag{24}
$$

It is clear from (24) a that the first two auto-terms are able to generate impulses which the cross terms cannot generate, and therefore, although the existence of cross terms has a certain influence on the detection performance, but the bi-component LFM signal still can be detected. This indicates that the scaled AFQ is also useful and powerful for detecting bi-component LFM signals. Moreover for an adequate value of k and matrix parameter Ω, the scaled AFQ benefits in cross-term reduction while maintaining a perfect time-frequency resolution with clear auto terms angle resolution.

5. Conclusion

Motivated by degree of freedom corresponding to the choice of a factor k in the fractional instantaneous auto-correlation and the extra degree of freedom present in QPFT, we proposed novel scaled AFQ. First, we studied the fundamental properties of the proposed distributions, including the time marginal, conjugate symmetry, non-linearity, time shift, frequency shift, frequency marginal, scaling, inverse and Moyal formula. Finally to show the of advantage of the theory, we provided the applications of the scaled AFQ in the detection of single-component and bi-component linear- frequency-modulated (LFM) signal.

Acknowledgements

This work is supported by Research project(JKSTIC/SRE/J/357-60) provided by DST Government of Jammu and Kashmir, India.

Author contributions

Both the authors contributed equally in the paper.

Conflict of interest

The authors declare no potential conflict of interests.

Classification

2000 Mathematics subject classification:

42C40; 81S30; 11R52; 44A35

Author details

Mohammad Younus Bhat[1*], Aamir Hamid Dar[1], Altaf Ahmad Bhat[2] and Deepak Kumar Jain[3]

1 Department of Mathematical Sciences, Islamic University of Science and Technology, Kashmir, India

2 University of Technology and Applied Sciences, Salalah, Oman

3 Madhav Institute of Technology and Science, Gwalior, India

*Address all correspondence to: gyounusg@gmail.com

References

[1] Castro LP, Haque MR, Murshed MM, Saitoh S, Tuan NM. Quadratic Fourier transforms. Annals of Functional Analysis. 2014;**5**(1):10-23

[2] Castro LP, Minh LT, Tuan NM. New convolutions for quadratic-phase Fourier integral operators and their applications. Mediterranean Journal of Mathematics. 2018;**15**(13). DOI: 10.1007/s00009-017-1063-y

[3] Bhat MY, Dar AH, Urynbassarova D, Urynbassarova A. Quadratic-phase wave packet transform. Optik - International Journal for Light and Electron Optics. Accepeted

[4] Shah FA, Nisar KS, Lone WZ, Tantary AY. Uncertainty principles for the quadratic-phase Fourier transforms. Mathematicsl Methods in the Applied Sciences. 2021. DOI: 10.1002/mma.7417

[5] Shah FA, Lone WZ, Tantary AY. Short-time quadratic-phase Fourier transform. Optik - International Journal for Light and Electron Optics. 2021. DOI: 10.1016/j.ijleo.2021.167689

[6] Shah FA, Teali AA. Quadratic-phase Wigner distribution. Optik - International Journal for Light and Electron Optics. 2021. DOI: 10.1016/j.ijleo.2021.168338

[7] Urynbassarova D, Li BZ, Tao R. The Wigner-Ville distribution in the linear canonical transform domain. IAENG International Journal of Applied Mathematics. 2016;**46**(4):559-563

[8] Urynbassarova D, Urynbassarova A, Al-Hussam E. The Wigner-Ville distribution based on the offset linear canonical transform domain. In: 2nd International Conference on Modelling, Simulation and Applied Mathematics. March 2017

[9] Bhat MY, Dar AH. Convolution and correlation theorems for Wigner-Ville distribution associated with the quaternion offset linear canonical transform. Signal Image and Video Processing

[10] Johnston JA. Wigner distribution and FM radar signal design. IEE Proceedings F: Radar and Signal Processing. 1989;**136**:81-88

[11] Wang MS, Chan AK, Chui CK. Linear frequency-modulated signal detection using radon-ambiguity transform. IEEE Transactions on Signal Processing. 1998;**46**:571-586

[12] Auslander L, Tolimieri R. Radar ambiguity functions and group theory. SIAM Journal on Mathematical Analysis. 1985;**16**:577-601

[13] Kutyniok G. Ambiguity functions, Wigner distributions and Cohen's class for LCA groups. Journal of Mathematical Analysis and Applications. 2003;**277**: 589-608

[14] Zhang, ZY, Levoy M. Wigner distributions and how they relate to the light field. In: Proc. IEEE International Conference Comput. Photography. 2009. pp. 1–10

[15] Qian S, Chen D. Joint time-frequency analysis. IEEE Signal Processing Magazine. 1999;**16**:52-67

[16] Cohen L. Time Frequency Analysis: Theory and Applications. Upper Saddle River, NJ: Prentice-Hall PTR; 1995

[17] Tao R, Deng B, Wang Y. Fractional Fourier Transform and its Applications. Beijing: Tsinghua University Press; 2009

[18] Shenoy RG, Parks TW. Wide-band ambiguity functions and affine Wigner distributions. Signal Processing. 1995; **41**(3):339-363

[19] Bastiaans MJ. Application of the Wigner distribution function in optics. Signal Processing. 1997;**375**:426

[20] Zhang ZC. Choi-williams distribution in linear canonical domains and its application in noisy LFM signals detection. Communications in Nonlinear Science and Numerical Simulation. 2020; **82**:105025

[21] Lu J, Oruklu E, Saniie J. Improved time-frequency distribution using singular value decomposition of Choi-Williams distribution. In: 2013 IEEE International Conference on Electro-Information Technology (EIT), Rapid City, SD, USA. 2013. pp. 1–4

[22] Choi HI, Williams WJ. Improved time-frequency representation of multicomponent signals using exponential kernels. IEEE Transactions on Acoustics, Speech, and Signal Processing. 1989;**37**(6):862-871

[23] Patti A, Williamson GA. Methods for classification of nocturnal migratory bird vocalizations using pseudo Wigner-Ville transform, In: 2013 IEEE International Conference on Acoustics, Speech and Signal Processing, Vancouver, BC, Canada. 2013. pp. 758–762

[24] Boashash B, O'Shea P. Polynomial wigner-ville distributions and their relationship to time-varying higher order spectra. IEEE Transactions on Signal Processing. 1994;**42**(1):216-220

[25] Stanković LJ, Stanković S. An analysis of instantaneous frequency representation using time-frequency distributions–generalized wigner distribution. IEEE Transactions on Signal Processing. 1995;**43**(2):549-552

[26] Stanković L. A method for time-frequency analysis. IEEE Transactions on Signal Processing. 1994;**42**(1):225-229

[27] Saulig N, Sucic V, Stanković S, Orivić I, Boashash B. Signal content estimation based on the short-term time-frequency Rényi entropy of the S-method time-frequency distribution. In: 2012 19th International Conference on Systems, Signals and Image Processing (IWSSIP), Vienna, Austria. 2012. pp. 354–357

[28] Zhang ZC, Jiang X, Qiang SZ, Sun A, Liang ZY, Shi X, et al. Scaled Wigner distribution using fractional instantaneous autocorrelation. Optik. 2021;**237**:166691

[29] Abolbashari M, Kim SM, Babaie G, Babaie J, Farahi F. Fractional bispectrum transform: definition and properties. IET Signal Processing. 2017;**11**(8):901-908

[30] Dar AH, Bhat MY. Scaled ambiguity function and scaled Wigner distribution for LCT signals. Optik - International Journal for Light and Electron Optics. 2022;**267**:169678

[31] Bhat MY, Dar AH. Scaled Wigner distribution in the offset linear canonical domain. Optik - International Journal for Light and Electron Optics. 2022;**262**: 169286

[32] Abe S, Sheridan JT. Optical operations on wave functions as the Abelian subgroups of the special affine Fourier transformation. Optics Letters. 1994;**19**(22):1801-1803

[33] Huo H. Uncertainty principles for the offset linear canonical transform. Circuits, Systems, and Signal Processing. 2019;**38**:395-406

[34] Huo H, Sun W, Xiao L. Uncertainty principles associated with the offset linear canonical transform. Mathematical Methods in the Applied Sciences. 2019;**42**:466-447

[35] Bhat MY, Dar AH. Octonion spectrum of 3D short-time LCT signals. arXiv:2202.00551 [eess.SP]. 2021

Section 2

Applications

Chapter 4

Analytical Expressions of Infinite Fourier Sine and Cosine Transform-Based Ramanujan Integrals $R_{S,C}(m, n)$ in Terms of Hypergeometric Series ${}_2F_3(\cdot)$

Showkat Ahmad Dar and M. Kamarujjama

Abstract

In this chapter, we obtain analytical expressions of infinite Fourier sine and cosine transform-based Ramanujan integrals,

$$\mathbf{R}_{S,C}(m, n) = \int_0^\infty \frac{x^m}{-1 + \exp(2\pi\sqrt{x})} \begin{matrix} \sin \\ \cos \end{matrix} (\pi n x) dx,$$

in an infinite series of hypergeometric functions ${}_2F_3(\cdot)$, using the hypergeometric technique. Also, we have given some generalizations of the Ramanujan's integrals $\mathbf{R}_{S,C}(m, n)$ in the form of integrals denoted by $\mathbf{I}^*_{S,C}(v, b, c, \lambda, y)$,$\mathbf{J}_{S,C}(v, b, c, \lambda, y)$, $\mathbf{K}_{S,C}(v, b, c, \lambda, y)$ and $\mathbf{I}_{S,C}(v, b, \lambda, y)$. These generalized definite integrals are expressed in terms of ordinary hypergeometric functions ${}_2F_3(\cdot)$, with suitable convergence conditions. Moreover, as applications of Ramanujan's integrals $\mathbf{R}_{S,C}(m, n)$, some closed form of infinite summation formulas involving hypergeometric functions ${}_1F_2$, ${}_2F_3(\cdot)$, and ${}_0F_1$ are derived.

Keywords: generalized hypergeometric function, infinite Fourier sine and cosine transforms, Ramanujan's integrals, Fox-Wright psi hypergeometric function, hypergeometric series

1. Introduction

Naturally, we call a function"special" when the function, just as the logarithm, the exponential and trigonometric functions (the elementary transcendental functions), belongs to the toolbox of the applied mathematician, the physicist, or the engineer. This branch of mathematics has a good history with great names such as Gauss, Euler, Fourier, Legendre, and Bessel. This chapter includes definitions, namely infinite Fourier sine and cosine transforms, Pochhammer's symbol and related results, generalized Gauss hypergeometric function and its special cases, Fox-Wright hypergeometric function and its convergence conditions, Hypergeometric form of elementary functions, Gauss-Legendre multiplication formula and infinite series decomposition identity. In the literature [1–6], the analytical expressions of the Fourier sine and cosine transforms of $x^{\upsilon-1}\backslash(\exp(bx)\pm 1)$ are available in terms of Riemann's zeta function, the Psi function (Digamma function), hyperbolic function and Beta function. The analytical solution of the following infinite Fourier sine and cosine transforms based Ramanujan integrals ([7], p. 85, eq. (49) last line):

$$\mathbf{R}_{S,C}(m,n)=\int_0^\infty \frac{x^m}{\{-1+\exp(2\pi\sqrt{x})\}}\begin{matrix}\sin\\ \cos\end{matrix}(\pi n x)dx, \tag{1}$$

are not given for all positive rational values of n and non-negative integral values of m.

2. Definitions and preliminaries

2.1 Fourier sine and cosine transforms

The infinite Fourier sine and cosine transforms of $g(x)$ over the interval $[0,\infty)$ are defined by

$$F_{S,C}(g(x);b)=\int_0^\infty g(x)\begin{matrix}\sin\\ \cos\end{matrix}(bx)dx=G_{S,C}(b),\ (b>0). \tag{2}$$

For example, if $y>0$, $0<Re(\upsilon)<2$ for Fourier sine transform of $x^{-\upsilon}$ and $y>0$, $0<Re(\upsilon)<1$ for Fourier cosine transform of $x^{-\upsilon}$, then the infinite Fourier sine and cosine transforms of $x^{-\upsilon}$ ([3], p. 68) are given by

$$\int_0^\infty x^{-\upsilon}\begin{matrix}\sin\\ \cos\end{matrix}(xy)dx=y^{\upsilon-1}\Gamma(1-\upsilon)\begin{matrix}\sin\\ \cos\end{matrix}\left(\frac{\upsilon\pi}{2}\right) \tag{3}$$

Further, if $b>0$, $-1<\Re(s)<1$ for Fourier sine transform and $b>0$, $0<\Re(s)<1$ for Fourier cosine transform, then the infinite Fourier sine and cosine transform of x^{s-1} are given by [3, 5, 8].

$$\int_0^\infty x^{s-1}\begin{matrix}\sin\\ \cos\end{matrix}(bx)dx=\frac{\Gamma(s)\begin{matrix}\sin\\ \cos\end{matrix}\left(\frac{\pi s}{2}\right)}{b^s}. \tag{4}$$

Moreover, if $\Re(\mu)>-2$ for Fourier sine transform and $\Re(\mu)>-1$ for Fourier cosine transform, then we can prove the following integral by using Maclaurin's expansion of $\exp(-ax^{\xi})$ and term by term integrating with the help of the result (4)

$$\int_0^\infty x^{\mu}\,exp\left(-ax^{\xi}\right)\begin{matrix}\sin\\ \cos\end{matrix}(xy)dx=y^{-\mu-1}\sum_{\ell=0}^{\infty}\left(-\frac{a}{y^{\xi}}\right)^{\ell}\frac{1}{\ell!}\Gamma(\mu+1+\xi\ell)\begin{matrix}\sin\\ \cos\end{matrix}\left\{\frac{\pi}{2}(\mu+\xi\ell)\right\}. \tag{5}$$

where $0<\xi<1$, $a>0$ and $y>0$. The conditions $\Re(\mu)>-2$ and $\Re(\mu)>-1$ stated in the integrals (5) follows from the theory of analytic continuation [5, 8]. We have also verified the conditions $\Re(\mu)>-2$ and $\Re(\mu)>-1$, using Wolfram Mathematica software.

2.2 Generalized gauss hypergeometric function

A natural generalization of the Gauss hypergeometric function ${}_2F_1(z)$ is the generalized hypergeometric function ${}_pF_q(z)$ with p numerator parameters $\alpha_1, \ldots, \alpha_p$ and q denominator parameters $\beta_1, \ldots, \beta_q$ defined by [9].

$$ {}_pF_q\left(\begin{matrix} \alpha_1, & \ldots, & \alpha_p; \\ \beta_1, & \ldots, & \beta_q; \end{matrix} \; z \right) = \sum_{n=0}^{\infty} \frac{(\alpha_1)_n \cdots (\alpha_p)_n}{(\beta_1)_n \cdots \left(\beta_q\right)_n} \frac{z^n}{n!}, \tag{6} $$

where $\alpha_j \in \mathbb{C}$ $(j = 1, \ldots, p)$, $\beta_j \in \mathbb{C}\backslash\mathbb{Z}_0^-$ $(j = 1, \ldots, q)$ and $p, q \in \mathbb{N}_0$. Then the hypergeometric ${}_pF_q(z)$ function in (6) converges absolutely for $|z| < \infty$ when $p \leq q$ and for $|z| < 1$ when $p = q + 1$. Furthermore, if we set,

$$ \omega := \left(\sum_{j=1}^{q} \beta_j - \sum_{j=1}^{p} \alpha_j \right), \tag{7} $$

it is known that when $p = q + 1$ the function ${}_pF_q(z)$ is absolutely convergent for $|z| = 1$ if $\Re(\omega) > 0$, conditionally convergent for $|z| = 1$ $(z \neq 1)$ if $-1 < \Re(\omega) < 0$ and divergent for $|z| = 1$ if $\Re(\omega) \leq -1$.

2.3 Hypergeometric form of elementary functions

The important special cases of ${}_pF_q(z)$ include (for example) the binomial series ${}_1F_0(z)$ given by [9].

$$ (1-z)^{-a} - {}_1F_0\left(\begin{matrix} a; \\ -; \end{matrix} \; z \right) = \sum_{n=0}^{\infty} \frac{(a)_n}{n!} z^n, \tag{8} $$

where $|z| < 1$, $a \in \mathbb{C}$.

Elementary trigonometric functions ([10], p. 44, eq. (9) and eq. (10)) are given by

$$ \cos z = {}_0F_1\left(\begin{matrix} -; \\ \frac{1}{2}; \end{matrix} \; \frac{-z^2}{4} \right), \tag{9} $$

$$ \sin z = z\,{}_0F_1\left(\begin{matrix} -; \\ \frac{3}{2}; \end{matrix} \; \frac{-z^2}{4} \right). \tag{10} $$

Lommel function ([10], p. 44, eq. (13)) is given by

$$ s_{\mu,\upsilon}(z) = \frac{z^{\mu+1}}{(\mu-\upsilon+1)(\mu+\upsilon+1)} {}_1F_2\left(\begin{matrix} & 1; & \\ \frac{\mu-\upsilon+3}{2}, & \frac{\mu+\upsilon+3}{2}; & \end{matrix} \; \frac{-z^2}{4} \right), \tag{11} $$

where $\mu \pm \upsilon \in \mathbb{C}\backslash\{-1, -3, -5, -7, \ldots\}$.

Struve function ([10], p. 44, eq. (16)) is given by

$$H_\upsilon(z) = \frac{2\left(\frac{z}{2}\right)^{\upsilon+1}}{\sqrt{\pi}\,\Gamma\left(\upsilon+\frac{3}{2}\right)}\,{}_1F_2\left(\begin{matrix} 1; \\ \frac{3}{2}, \upsilon+\frac{3}{2}; \end{matrix}\ \frac{-z^2}{4}\right). \tag{12}$$

Modified Struve function ([10], p. 45, eq. (17)) is given by

$$L_\upsilon(z) = \frac{2\left(\frac{z}{2}\right)^{\upsilon+1}}{\sqrt{\pi}\,\Gamma\left(\upsilon+\frac{3}{2}\right)}\,{}_1F_2\left(\begin{matrix} 1; \\ \frac{3}{2}, \upsilon+\frac{3}{2}; \end{matrix}\ \frac{z^2}{4}\right). \tag{13}$$

2.4 Pochhammer's symbol

Here $(\lambda)_\upsilon$ $(\lambda, \upsilon \in \mathbb{C})$ denotes the Pochhammer's symbol (or the shifted factorial, since $(1)_n = n!$) is defined, in general, by [10].

$$(\lambda)_\upsilon := \frac{\Gamma(\lambda+\upsilon)}{\Gamma(\lambda)} = \begin{cases} 1, & (\upsilon = 0; \lambda \in \mathbb{C}\backslash\{0\}) \\ \lambda(\lambda+1)\ldots(\lambda+n-1), & (\upsilon = n \in \mathbb{N}; \lambda \in \mathbb{C}). \end{cases} \tag{14}$$

Algebraic property of Pochhammer symbol:

$$(\lambda)_{m+n} = (\lambda)_m(\lambda+m)_n = (\lambda)_n(\lambda+n)_m. \tag{15}$$

2.5 Gauss-Legendre multiplication formula

For every positive integer m ([10], p. 22, eq. (26)), we have

$$(\lambda)_{mn} = m^{mn}\prod_{j=1}^{m}\left(\frac{\lambda+j-1}{m}\right)_n \qquad ; m \in \mathbb{N}, n \in \mathbb{N}_0. \tag{16}$$

From the above result (16) with $\lambda = mz$, it can be proved that

$$\Gamma(mz) = (2\pi)^{\frac{(1-m)}{2}} m^{mz-\frac{1}{2}}\prod_{j=1}^{m}\Gamma\left(z+\frac{j-1}{m}\right), \tag{17}$$

where $z \neq 0, -\frac{1}{m}, -\frac{2}{m}, \ldots; m \in \mathbb{N}$.
The eq. (17) is known as Gauss-Legendre multiplication formula for Gamma function.

2.6 Legendre's duplication formula

When we put $m = 2$ in the eq. (17), we get

$$\sqrt{\pi}\Gamma(2z) = 2^{2z-1}\Gamma(z)\Gamma\left(z+\frac{1}{2}\right), \qquad 2z \in \mathbb{C}\backslash\mathbb{Z}_0^-, \tag{18}$$

which is known as Legendre's duplication formula.

2.7 Infinite series decomposition identity

An infinite series decomposition identity ([11], p. 193, eq. (8)) is given by

$$\sum_{\ell=0}^{\infty}\Omega(\ell)=\sum_{j=0}^{N-1}\left\{\sum_{\ell=0}^{\infty}\Omega(N\ell+j)\right\}, \tag{19}$$

where N is an arbitrary positive integer. Put $N = 4$ in the above eq. (19), we get

$$\sum_{\ell=0}^{\infty}\Omega(\ell)=\sum_{j=0}^{3}\left\{\sum_{\ell=0}^{\infty}\Omega(4\ell+j)\right\}, \tag{20}$$

$$=\sum_{\ell=0}^{\infty}\Omega(4\ell)+\sum_{\ell=0}^{\infty}\Omega(4\ell+1)+\sum_{\ell=0}^{\infty}\Omega(4\ell+2)+\sum_{\ell=0}^{\infty}\Omega(4\ell+3), \tag{21}$$

provided that all involved infinite series are absolutely convergent.

2.8 Fox-Wright psi function of one variable

A natural generalization of the hypergeometric function ${}_pF_q(z)$ is the Fox-Wright psi function of one variable with p pairs of numerator parameters $(\alpha_1, A_1), \ldots, (\alpha_p, A_p)$ and q pairs of denominator parameters $(\beta_1, B_1), \ldots, \left(\beta_q, B_q\right)$, defined by [12, 13].

$${}_p\Psi_q\left[\begin{matrix}(\alpha_1,\ A_1),\ \ldots,\ (\alpha_p,\ A_p); \\ (\beta_1,\ B_1),\ \ldots,\ \left(\beta_q,\ B_q\right);\end{matrix}\ z\right]=\sum_{k=0}^{\infty}\frac{\Gamma(\alpha_1+kA_1)\ldots\Gamma\left(\alpha_p+kA_p\right)}{\Gamma(\beta_1+kB_1)\ldots\Gamma\left(\beta_q+kB_q\right)}\frac{z^k}{k!}, \tag{22}$$

$$=\frac{1}{2\pi\rho}\int_L\frac{\Gamma(\zeta)\prod_{i=1}^{p}\Gamma(\alpha_i-A_i\zeta)}{\prod_{j=1}^{q}\Gamma\left(\beta_j-B_j\zeta\right)}(-z)^{-\zeta}d\zeta, \tag{23}$$

where $\rho=\sqrt{-1}, z\in\mathbb{C}$; parameters $\alpha_i, \beta_j\in\mathbb{C}$; coefficients $A_i, B_j\in\mathbb{R}=(-\infty, +\infty)$ in case of series (22) (or $A_i, B_j\in\mathbb{R}_+=(0, +\infty)$ in case of contour integral (23)), $A_i\neq 0\ (i=1, 2, \ldots, p), B_j\neq 0\ (j=1, 2, \ldots, q)$. In eq. (22), the parameters α_i, β_j and coefficients A_i, B_j are adjusted in such a way that the product of Gamma functions in numerator and denominator should be well defined.

Suppose:

$$\Delta^*=\left(\sum_{j=1}^{q}B_j-\sum_{i=1}^{p}A_i\right), \tag{24}$$

$$\delta^*=\left(\prod_{i=1}^{p}|A_i|^{-A_i}\right)\left(\prod_{j=1}^{q}|B_j|^{B_j}\right), \tag{25}$$

$$\mu^*=\sum_{j=1}^{q}\beta_j-\sum_{i=1}^{p}\alpha_i+\left(\frac{p-q}{2}\right), \tag{26}$$

and

$$\sigma^* = \left(1 + A_1 + \ldots + A_p\right) - \left(B_1 + \ldots + B_q\right) = 1 - \Delta^*. \tag{27}$$

Then we have the following convergence conditions of (22) or (23).

Case (1): When contour (L) is a left loop beginning and ending at $-\infty$, then ${}_p\Psi_q[\cdot]$ given by (22) or (23) holds the following convergence conditions.

i. When $\Delta^* > -1$, $0 < |z| < \infty$, $z \neq 0$.

ii. When $\Delta^* = -1$, $0 < |z| < \delta^*$.

iii. When $\Delta^* = -1, |z| = \delta^*$, and $\Re(\mu^*) > \frac{1}{2}$.

Case (2): When contour (L) is a right loop beginning and ending at $+\infty$, then ${}_p\Psi_q[\cdot]$ given by (22) or (23) holds the following convergence conditions.

i. When $\Delta^* < -1$, $0 < |z| < \infty, z \neq 0$.

ii. When $\Delta^* = -1$, $|z| > \delta^*$.

iii. When $\Delta^* = -1, |z| = \delta^*$, and $\Re(\mu^*) > \frac{1}{2}$.

Case (3): When contour (L) is starting from $\gamma - i\infty$ and ending at $\gamma + i\infty$ where $\gamma \in \mathbb{R} = (-\infty, +\infty)$, then ${}_p\Psi_q[\cdot]$ is also convergent under the following conditions.

i. When $\sigma^* > 0$, $|arg(-z)| < \frac{\pi}{2}\sigma^*$, $0 < |z| < \infty, z \neq 0$.

ii. When $\sigma^* = 0, arg(-z) = 0, 0 < |z| < \infty, z \neq 0$ such that $-\gamma\Delta^* + \Re(\mu^*) > \frac{1}{2} + \gamma$.

iii. When $\gamma = 0$, $\sigma^* = 0, arg(-z) = 0$, $0 < |z| < \infty, z \neq 0$, such that $\Re(\mu^*) > \frac{1}{2}$.

In the available literature [7, 14–18] on Ramanujan's Mathematics, the analytical expression of Ramanujan's integrals $\mathbf{R}_{S,C}(m, n)$ are not given. Therefore, the main object of this chapter is to evaluate the representation of $\mathbf{R}_{S,C}(m, n)$ in an ordinary hypergeometric function ${}_2F_3(\cdot)$. Also, our contribution to Ramanujan's Mathematics is determined by the result in [19, 20]. Here in this chapter, we generalize Ramanujan's integrals $\mathbf{R}_{S,C}(m, n)$ in the following forms:

$$\mathbf{I}^*_{S,C}(\upsilon, b, c, \lambda, y) = \sum_{k=0}^{\infty} \frac{\Theta(k)}{k!} \int_0^\infty x^{\upsilon-1} e^{-(\lambda b + ck)\sqrt{x}} \begin{matrix} \sin \\ \cos \end{matrix} (xy) dx,$$

$$\mathbf{J}_{S,C}(\upsilon, b, c, \lambda, y) = \int_0^\infty x^{\upsilon-1} e^{-b\lambda\sqrt{x}} {}_r\Psi_s \left[\begin{matrix} (\alpha_1, A_1), \ldots, (\alpha_r, A_r); \\ (\beta_1, B_1), \ldots, (\beta_s, B_s); \end{matrix} \quad e^{-c\sqrt{x}} \right] \begin{matrix} \sin \\ \cos \end{matrix} (xy) dx,$$

$$\mathbf{K}_{S,C}(\upsilon, b, c, \lambda, y) = \int_0^\infty x^{\upsilon-1} e^{-b\lambda\sqrt{x}} {}_rF_s \left(\begin{matrix} \alpha_1, \ldots, \alpha_r; \\ \beta_1, \ldots, \beta_s; \end{matrix} \quad e^{-c\sqrt{x}} \right) \begin{matrix} \sin \\ \cos \end{matrix} (xy) dx,$$

$$\mathbf{I}_{S,C}(\upsilon, b, \lambda, y) = \int_0^\infty x^{\upsilon-1} \left\{-1 + \exp\left(b\sqrt{x}\right)\right\}^{-\lambda} \begin{matrix} \sin \\ \cos \end{matrix} (xy) dx,$$

where $\{\Theta(k)\}_{k=0}^{\infty}$ is a fixed sequence of the arbitrary real or complex numbers. Moreover, we also show how the main general theorem given below applies to obtaining new interesting results by suitable adjustments in parameters and variables.

3. Ramanujan's integrals

The analytical solution of the following integral of Ramanujan ([7], p. 85, eq. (49) last line):

$$\mathbf{R}_C(m, n) = \int_0^{\infty} x^m \frac{\cos(\pi n x)}{\{-1 + \exp(2\pi\sqrt{x})\}} dx, \tag{28}$$

is not given for all positive rational values of n and non-negative integral values of m.

For particular values of m and n in Ramanujan's integral $\mathbf{R}_C(m, n)$, the following three integrals are given by ([7], p. 86, eq. (50)):

$$\mathbf{R}_C(1, 1/2) = \int_0^{\infty} \frac{x\cos\left(\frac{\pi x}{2}\right)}{\{-1 + \exp(2\pi\sqrt{x})\}} dx = \frac{13 - 4\pi}{8\pi^2}, \tag{29}$$

$$\mathbf{R}_C(1, 2) = \int_0^{\infty} \frac{x\cos(2\pi x)}{\{-1 + \exp(2\pi\sqrt{x})\}} dx = \frac{1}{64}\left(\frac{1}{2} - \frac{3}{\pi} + \frac{5}{\pi^2}\right), \tag{30}$$

$$\mathbf{R}_C(2, 2) = \int_0^{\infty} \frac{x^2\cos(2\pi x)}{\{-1 + \exp(2\pi\sqrt{x})\}} dx = \frac{1}{256}\left(1 - \frac{5}{\pi} + \frac{5}{\pi^2}\right). \tag{31}$$

The following theorem is proved by Ramanujan ([7], p. 76, 77, eq. (10 and 10′)).

Theorem 1.3.1. *Let n be real and positive. Then if*

$$R_C(0, n) = \Phi(n) = \int_0^{\infty} \frac{\cos(\pi n x)}{\{-1 + \exp(2\pi\sqrt{x})\}} dx, \tag{32}$$

and

$$\Upsilon(n) - \frac{1}{2\pi n} = \int_0^{\infty} \frac{\sin(\pi n x)}{\{-1 + \exp(2\pi\sqrt{x})\}} dx = R_S(0, n), \tag{33}$$

then

$$\boldsymbol{R}_C(0, n) = \Phi(n) = \frac{1}{n}\sqrt{\left(\frac{2}{n}\right)}\,\Upsilon\left(\frac{1}{n}\right) - \Upsilon(n), \tag{34}$$

and

$$\Upsilon(n) = \frac{1}{n}\sqrt{\left(\frac{2}{n}\right)}\,\Phi\left(\frac{1}{n}\right) + \Phi(n), \tag{35}$$

For particular values of n, some values of Ramanujan's integral ([7], p. 85 (eq. 48)) are given below

$$\mathbf{R}_C(0, 1) = \Phi(1) = \int_0^\infty \frac{\cos(\pi x)}{\{-1 + \exp(2\pi\sqrt{x})\}} dx = \frac{2 - \sqrt{2}}{8}, \tag{36}$$

$$\mathbf{R}_C(0, 2) = \Phi(2) = \int_0^\infty \frac{\cos(2\pi x)}{\{-1 + \exp(2\pi\sqrt{x})\}} dx = \frac{1}{16}, \tag{37}$$

$$\mathbf{R}_C(0, 4) = \Phi(4) = \int_0^\infty \frac{\cos(4\pi x)}{\{-1 + \exp(2\pi\sqrt{x})\}} dx = \frac{3 - \sqrt{2}}{32}, \tag{38}$$

$$\mathbf{R}_C(0, 6) = \Phi(6) = \int_0^\infty \frac{\cos(6\pi x)}{\{-1 + \exp(2\pi\sqrt{x})\}} dx = \frac{13 - 4\sqrt{3}}{144}, \tag{39}$$

$$\mathbf{R}_C(0, 1/2) = \Phi\left(\frac{1}{2}\right) = \int_0^\infty \frac{\cos\left(\frac{\pi x}{2}\right)}{\{-1 + \exp(2\pi\sqrt{x})\}} dx = \frac{1}{4\pi}, \tag{40}$$

$$\mathbf{R}_C(0, 2/5) = \Phi\left(\frac{2}{5}\right) = \int_0^\infty \frac{\cos\left(\frac{2\pi x}{5}\right)}{\{-1 + \exp(2\pi\sqrt{x})\}} dx = \frac{8 - 3\sqrt{5}}{16}. \tag{41}$$

By calculation of (7) and (9), we get the following infinite fourier sine transform of $-1 + \exp(2\pi\sqrt{x})$

$$\int_0^\infty \frac{\sin(\pi n x)}{\{-1 + \exp(2\pi\sqrt{x})\}} dx = \frac{1}{n}\sqrt{\left(\frac{2}{n}\right)}\,\Phi\left(\frac{1}{n}\right) + \Phi(n) - \frac{1}{2\pi n}. \tag{42}$$

For special values of $n = 1{,}2,\frac{1}{2}$ in the above eq. (16) and using $\Phi(1)$,$\Phi(2)$ and $\Phi\left(\frac{1}{2}\right)$, we get after simplification the following three results:

$$\mathbf{R}_S(0, 1) = \int_0^\infty \frac{\sin(\pi x)}{\{-1 + \exp(2\pi\sqrt{x})\}} dx = \frac{\pi\sqrt{2} - 4}{8\pi}, \tag{43}$$

$$\mathbf{R}_S(0, 2) = \int_0^\infty \frac{\sin(2\pi x)}{\{-1 + \exp(2\pi\sqrt{x})\}} dx = \frac{\pi - 2}{16\pi}, \tag{44}$$

$$\mathbf{R}_S(0, 1/2) = \int_0^\infty \frac{\sin\left(\frac{\pi x}{2}\right)}{\{-1 + \exp(2\pi\sqrt{x})\}} dx = \frac{\pi - 3}{4\pi}. \tag{45}$$

4. Main general theorems on infinite Fourier sine and cosine transform

In this section, we give some generalizations of the infinite Fourier sine and cosine transform-based Ramanujan integrals $\mathbf{R}_{S,C}(\cdot)$ in the form of infinite series of hypergeometric functions ${}_2F_3(\cdot)$. Moreover, we denote these generalizations by $\mathbf{I}^*_{S,C}(\cdot)$, $\mathbf{J}_{S,C}(\cdot)$, $\mathbf{K}_{S,C}(\cdot)$ and $\mathbf{I}_{S,C}(\cdot)$ [21, 22].

Theorem 1.4.1. *Suppose* $\{\Theta(k)\}_{k=0}^\infty$ *is a fixed sequence of arbitrary real or complex numbers and satisfy the conditions* $\Re(v) > -1, c > 0, y > 0; \lambda > 0, b > 0$ (*or* $\lambda < 0$, $b < 0$).

then we have

$$\mathbf{I}^*_{S,C}(v, b, c, \lambda, y) = \sum_{k=0}^\infty \frac{\Theta(k)}{k!}\int_0^\infty x^{v-1} e^{-(\lambda b + ck)\sqrt{x}} \begin{matrix}\sin\\ \cos\end{matrix}(xy)dx, \tag{46}$$

$$= y^{-\upsilon} \sum_{k=0}^{\infty} \frac{\Theta(k)}{k!} \sum_{\ell=0}^{\infty} \frac{(-1)^{\ell} (\lambda b + ck)^{\ell} \Gamma\left(\upsilon + \frac{\ell}{2}\right)}{y^{\frac{\ell}{2}} \ell!} \begin{matrix} \sin \\ \cos \end{matrix} \left(\frac{\upsilon\pi}{2} + \frac{\ell\pi}{4}\right), \tag{47}$$

Now replacing ℓ by $4\ell + j$, after simplification we get

$$\mathbf{I}^{*}_{S,C}(\upsilon, b, c, \lambda, y) = y^{-\upsilon} \sum_{k=0}^{\infty} \frac{\Theta(k)}{k!} \sum_{j=0}^{3} \frac{(-1)^{j} (\lambda b + ck)^{j} \Gamma\left(\upsilon + \frac{j}{2}\right)}{y^{\frac{j}{2}} j!} \begin{matrix} \sin \\ \cos \end{matrix} \left(\frac{\upsilon\pi}{2} + \frac{j\pi}{4}\right) \times {}_2F_3\left(\begin{matrix} \Delta\left(2; \frac{2\upsilon + j}{2}\right); \\ \Delta^{*}(4; 1+j); \end{matrix} \frac{-1}{64y^2} \left\{ \frac{(\lambda b)\left(\frac{\lambda b + c}{c}\right)_k}{\left(\frac{\lambda b}{c}\right)_k} \right\}^4 \right), \tag{48}$$

$$= y^{-\upsilon} \sum_{k=0}^{\infty} \frac{\Theta(k)}{k!} \sum_{j=0}^{3} \frac{(-1)^{j} \Gamma\left(\upsilon + \frac{j}{2}\right)}{j!} \begin{matrix} \sin \\ \cos \end{matrix} \left(\frac{\upsilon\pi}{2} + \frac{j\pi}{4}\right) \left(\frac{\lambda b}{\sqrt{y}}\right)^{j} \times \left\{ \frac{\left(\frac{\lambda b + c}{c}\right)_k}{\left(\frac{\lambda b}{c}\right)_k} \right\}^{j} {}_2F_3\left(\begin{matrix} \Delta\left(2; \frac{2\upsilon + j}{2}\right); \\ \Delta^{*}(4; 1+j); \end{matrix} \frac{-1}{64y^2} \left\{ \frac{(\lambda b)\left(\frac{\lambda b + c}{c}\right)_k}{\left(\frac{\lambda b}{c}\right)_k} \right\}^4 \right), \tag{49}$$

$$\begin{aligned}
&= \frac{\Gamma(\upsilon) \begin{matrix} \sin \\ \cos \end{matrix} \left(\frac{\upsilon\pi}{2}\right)}{y^{\upsilon}} \sum_{k=0}^{\infty} \frac{\Theta(k)}{k!} {}_2F_3\left(\begin{matrix} \frac{\upsilon}{2}, \frac{\upsilon+1}{2}; \\ \frac{1}{4}, \frac{1}{2}, \frac{3}{4}; \end{matrix} \frac{-1}{64y^2} \left\{ \frac{(\lambda b)\left(\frac{\lambda b + c}{c}\right)_k}{\left(\frac{\lambda b}{c}\right)_k} \right\}^4 \right) \\
&- \frac{(\lambda b)\Gamma\left(\upsilon + \frac{1}{2}\right) \begin{matrix} \sin \\ \cos \end{matrix} \left(\frac{\upsilon\pi}{2} + \frac{\pi}{4}\right)}{y^{\upsilon + \frac{1}{2}}} \sum_{k-0}^{\infty} \frac{\Theta(k)}{k!} \left\{ \frac{\left(\frac{\lambda b + c}{c}\right)_k}{\left(\frac{\lambda b}{c}\right)_k} \right\} \\
&\times {}_2F_3\left(\begin{matrix} \frac{2\upsilon+1}{4}, \frac{2\upsilon+3}{4}; \\ \frac{1}{2}, \frac{3}{4}, \frac{5}{4} \quad ; \end{matrix} \frac{-1}{64y^2} \left\{ \frac{(\lambda b)\left(\frac{\lambda b + c}{c}\right)_k}{\left(\frac{\lambda b}{c}\right)_k} \right\}^4 \right) \\
&+ \frac{(\lambda b)^2 \Gamma(\upsilon + 1) \begin{matrix} \sin \\ \cos \end{matrix} \left(\frac{\upsilon\pi}{2}\right)}{2y^{\upsilon+1}} \sum_{k=0}^{\infty} \frac{\Theta(k)}{k!} \left\{ \frac{\left(\frac{\lambda b + c}{c}\right)_k}{\left(\frac{\lambda b}{c}\right)_k} \right\}^2 \\
&\times {}_2F_3\left(\begin{matrix} \frac{\upsilon+1}{2}, \frac{\upsilon+2}{2}; \\ \frac{3}{4}, \frac{5}{4}, \frac{3}{2} \quad ; \end{matrix} \frac{-1}{64y^2} \left\{ \frac{(\lambda b)\left(\frac{\lambda b + c}{c}\right)_k}{\left(\frac{\lambda b}{c}\right)_k} \right\}^4 \right) \\
&- \frac{(\lambda b)^3 \Gamma\left(\upsilon + \frac{3}{2}\right) \begin{matrix} \sin \\ \cos \end{matrix} \left(\frac{\upsilon\pi}{2} + \frac{\pi}{4}\right)}{6y^{\upsilon + \frac{3}{2}}} \sum_{k=0}^{\infty} \frac{\Theta(k)}{k!} \left\{ \frac{\left(\frac{\lambda b + c}{c}\right)_k}{\left(\frac{\lambda b}{c}\right)_k} \right\}^3 \\
&\times {}_2F_3\left(\begin{matrix} \frac{2\upsilon+3}{4}, \frac{2\upsilon+5}{4}; \\ \frac{5}{4}, \frac{3}{2}, \frac{7}{4} \quad ; \end{matrix} \frac{-1}{64y^2} \left\{ \frac{(\lambda b)\left(\frac{\lambda b + c}{c}\right)_k}{\left(\frac{\lambda b}{c}\right)_k} \right\}^4 \right).
\end{aligned} \tag{50}$$

Our result (30)or (31) or (32) is convergent in view of the convergence condition of ${}_pF_q(\cdot)$ series, when $p \leq q$, and $\forall |z| < \infty$.

Proof: The result (29) is obtained by the application of the integral (49) [with substitutions $\mu = \upsilon - 1, a = \lambda b + ck, \xi = \frac{1}{2}$] in the R.H.S. of eq. (28). Also, we calculate the results (30) to (32) by using the infinite series decomposition identity (20) and (21) and algebraic properties of Pochhammer's symbols.

4.1 Analytical expressions of infinite Fourier sine and cosine transforms

Theorem 1.4.2. *Analytical expressions of the infinite Fourier sine and cosine transforms of* $x^{\upsilon-1}e^{-b\lambda\sqrt{x}}{}_r\Psi_s[\cdot]$ *holds true for* $\Re(\upsilon) > -1; c > 0, y > 0; \lambda > 0, b > 0$ *(or* $\lambda < 0$, $b < 0$*) then we have*

$$J_{S,C}(\upsilon, b, c, \lambda, y) = \int_0^\infty x^{\upsilon-1}e^{-b\lambda\sqrt{x}}{}_r\Psi_s\left[\begin{array}{l}(\alpha_1,\ A_1),\ \dots,\ (\alpha_r,\ A_r);\\(\beta_1,\ B_1),\ \dots,\ (\beta_s,\ B_s);\end{array} e^{-c\sqrt{x}}\right]\begin{array}{c}\sin\\\cos\end{array}(xy)dx, \tag{51}$$

$$= y^{-\upsilon}\sum_{k=0}^{\infty}\frac{\Gamma(\alpha_1+kA_1)\dots\Gamma(\alpha_r+kA_r)}{\Gamma(\beta_1+kB_1)\dots\Gamma(\beta_s+kB_s)k!}\sum_{j=0}^{3}\frac{(-1)^j(\lambda b+ck)^j\,\Gamma\left(\upsilon+\frac{j}{2}\right)}{y^{\frac{j}{2}}j!}\times$$
$$\times\begin{array}{c}\sin\\\cos\end{array}\left(\frac{\upsilon\pi}{2}+\frac{j\pi}{4}\right){}_2F_3\left(\begin{array}{l}\Delta\left(2;\dfrac{2\upsilon+j}{2}\right);\\ \\ \Delta^*(4;1+j);\end{array}\ \frac{-1}{64y^2}\left\{\frac{(\lambda b)\left(\frac{\lambda b+c}{c}\right)_k}{\left(\frac{\lambda b}{c}\right)_k}\right\}^4\right), \tag{52}$$

Here the parameters $\alpha_i, \beta_j \in \mathbb{C}$ and coefficients $A_i, B_j \in \mathbb{R} = (-\infty, +\infty); A_i \neq 0\ (i = 1, 2, \dots, r), B_j \neq 0\ (j = 1, 2, \dots, s)$ and ${}_r\Psi_s[\cdot]$ is the Fox-Wright psi function of one variable subject to suitable convergence conditions derived from conditions discussed in case (1) or case (2) or case (3) of the function ${}_p\Psi_q[\cdot]$ given by (22) and (23). When N is positive integer then $\Delta(N;\lambda)$ denotes the array of N parameters given by $\frac{\lambda}{N}, \frac{\lambda+1}{N}, \dots, \frac{\lambda+N-1}{N}$. When N and j are independent variables then the notation $\Delta(N;j+1)$ denotes the set of N parameters given by $\frac{j+1}{N}, \frac{j+2}{N}, \dots, \frac{j+N}{N}$. When j is dependent variable that is $j = 0,1,2,3, \dots, N-1$, then the asterisk in $\Delta^*(N;j+1)$ represents the fact that the (denominator) parameters $\frac{N}{N}$ is always omitted (due to the need of factorial in denominator in the power series form of hypergeometric function) so that the set $\Delta^*(N;j+1)$ obviously contains only $(N-1)$ parameters ([10], Chap. 3, p. 214).

Proof: Let us consider $\Theta(k) = \frac{\Gamma(\alpha_1+kA_1)\dots\Gamma(\alpha_r+kA_r)}{\Gamma(\beta_1+kB_1)\dots\Gamma(\beta_s+kB_s)}$ $(k = 0,1,2,3, \dots)$ in the eqs. (28) and (30), then after evaluation we get integral expressions (33) involving the Fox-Wright Psi function in the form of infinite series of an ordinary ${}_2F_3$ hypergeometric function (34).

Theorem 1.4.3. *Analytical expressions of the infinite Fourier sine and cosine transforms of* $x^{\upsilon-1}e^{-b\lambda\sqrt{x}}{}_rF_s(\cdot)$ *holds true for* $\Re(\upsilon) > -1; c > 0, y > 0; \lambda > 0, b > 0$ *(or* $\lambda < 0, b < 0$*) then we have*

$$\boldsymbol{K}_{S,C}(\upsilon, b, c, \lambda, y) = \int_0^\infty x^{\upsilon-1}e^{-b\lambda\sqrt{x}}{}_rF_s\left(\begin{array}{l}\alpha_1,\ \dots,\ \alpha_r;\\\beta_1,\ \dots,\ \beta_s;\end{array} e^{-c\sqrt{x}}\right)\begin{array}{c}\sin\\\cos\end{array}(xy)dx, \tag{53}$$

$$= y^{-\upsilon} \sum_{k=0}^{\infty} \frac{(\alpha_1)_k \dots (\alpha_r)_k}{(\beta_1)_k \dots (\beta_s)_k \, k!} \sum_{j=0}^{3} \frac{(-1)^j (\lambda b + ck)^j \, \Gamma\left(\upsilon + \frac{j}{2}\right)}{y^{\frac{j}{2}} j!} \begin{matrix} \sin \\ \cos \end{matrix} \left(\frac{\upsilon\pi}{2} + \frac{j\pi}{4}\right)$$
$$\times {}_2F_3 \left(\begin{matrix} \Delta\left(2; \frac{2\upsilon + j}{2}\right); \\ \Delta^*(4; 1+j); \end{matrix} \; \frac{-1}{64y^2} \left\{ \frac{(\lambda b)\left(\frac{\lambda b + c}{c}\right)_k}{\left(\frac{\lambda b}{c}\right)_k} \right\}^4 \right), \tag{54}$$

where the parameters $\alpha_i, \beta_j \in \mathbb{C}$ $(i = 1, 2, \dots, r), (j = 1, 2, \dots, s)$ and $r \leq s + 1$.

Proof: If $A_1 = \dots = A_r = B_1 = \dots = B_s = 1$ in (33) and (34), then we get the above integral expressions involving generalized hypergeometric function in the form of infinite series of an ordinary ${}_2F_3$ hypergeometric function (36).

Corollary 1.4.4. *An infinite Fourier sine and cosine transforms of* $x^{\upsilon-1}\{-1 + \exp(b\sqrt{x})\}^{-\lambda}$ *holds true for* $\Re(\upsilon) > -1; \lambda > 0, b > 0$ *and* $y > 0$ *then we have*

$$\boldsymbol{I}_{S,C}(\upsilon, b, \lambda, y) = \int_0^{\infty} \frac{x^{\upsilon-1}}{\{-1 + \exp(b\sqrt{x})\}^{\lambda}} \begin{matrix} \sin \\ \cos \end{matrix} (xy)dx, \tag{55}$$

$$= y^{-\upsilon} \sum_{k=0}^{\infty} \frac{(\lambda)_k}{k!} \sum_{\ell=0}^{\infty} \frac{(-1)^{\ell} (\lambda b + bk)^{\ell} \Gamma\left(\upsilon + \frac{\ell}{2}\right)}{y^{\frac{\ell}{2}} \ell!} \begin{matrix} \sin \\ \cos \end{matrix} \left(\frac{\upsilon\pi}{2} + \frac{\ell\pi}{4}\right), \tag{56}$$

$$= y^{-\upsilon} \sum_{k=0}^{\infty} \frac{(\lambda)_k}{k!} \sum_{j=0}^{3} \frac{(-1)^j (\lambda b + bk)^j \, \Gamma\left(\upsilon + \frac{j}{2}\right)}{y^{\frac{j}{2}} j!} \begin{matrix} \sin \\ \cos \end{matrix} \left(\frac{\upsilon\pi}{2} + \frac{j\pi}{4}\right) \times$$
$$\times {}_2F_3 \left(\begin{matrix} \Delta\left(2; \frac{2\upsilon + j}{2}\right); \\ \Delta^*(4; 1+j); \end{matrix} \; \frac{-1}{64y^2} \left\{ \frac{(\lambda b)(\lambda + 1)_k}{(\lambda)_k} \right\}^4 \right), \tag{57}$$

$$= \begin{aligned} & \frac{\Gamma(\upsilon) \begin{matrix} \sin \\ \cos \end{matrix} \left(\frac{\upsilon\pi}{2}\right)}{y^{\upsilon}} \sum_{k=0}^{\infty} \frac{(\lambda)_k}{k!} {}_2F_3 \left(\begin{matrix} \frac{\upsilon}{2}, \frac{\upsilon+1}{2}; \\ \frac{1}{4}, \frac{1}{2}, \frac{3}{4}; \end{matrix} \; \frac{-1}{64y^2} \left\{ \frac{(\lambda b)(\lambda + 1)_k}{(\lambda)_k} \right\}^4 \right) \\ & - \frac{(\lambda b)\Gamma\left(\upsilon + \frac{1}{2}\right) \begin{matrix} \sin \\ \cos \end{matrix} \left(\frac{\upsilon\pi}{2} + \frac{\pi}{4}\right)}{y^{\upsilon + \frac{1}{2}}} \sum_{k=0}^{\infty} \frac{(\lambda + 1)_k}{k!} {}_2F_3 \left(\begin{matrix} \frac{2\upsilon+1}{4}, \frac{2\upsilon+3}{4}; \\ \frac{1}{2}, \frac{3}{4}, \frac{5}{4}; \end{matrix} \; \frac{-1}{64y^2} \left\{ \frac{(\lambda b)(\lambda + 1)_k}{(\lambda)_k} \right\}^4 \right) \\ & + \frac{(\lambda b)^2 \Gamma(\upsilon + 1) \begin{matrix} \sin \\ \cos \end{matrix} \left(\frac{\upsilon\pi}{2}\right)}{2y^{\upsilon+1}} \sum_{k=0}^{\infty} \frac{\{(\lambda + 1)_k\}^2}{(\lambda)_k \, k!} {}_2F_3 \left(\begin{matrix} \frac{\upsilon+1}{2}, \frac{\upsilon+2}{2}; \\ \frac{3}{4}, \frac{5}{4}, \frac{3}{2}; \end{matrix} \; \frac{-1}{64y^2} \left\{ \frac{(\lambda b)(\lambda + 1)_k}{(\lambda)_k} \right\}^4 \right) \\ & - \frac{(\lambda b)^3 \Gamma\left(\upsilon + \frac{3}{2}\right) \begin{matrix} \sin \\ \cos \end{matrix} \left(\frac{\upsilon\pi}{2} + \frac{\pi}{4}\right)}{6y^{\upsilon + \frac{3}{2}}} \sum_{k=0}^{\infty} \frac{\{(\lambda + 1)_k\}^3}{k! \{(\lambda)_k\}^2} {}_2F_3 \left(\begin{matrix} \frac{2\upsilon+3}{4}, \frac{2\upsilon+5}{4}; \\ \frac{5}{4}, \frac{3}{2}, \frac{7}{4}; \end{matrix} \; \frac{-1}{64y^2} \left\{ \frac{(\lambda b)(\lambda + 1)_k}{(\lambda)_k} \right\}^4 \right), \end{aligned} \tag{58}$$

Proof: If we consider $\Theta(k) = (\lambda)_k$ and $c = b$ in (28), which yields

$$\mathbf{I}_{S,C}(\upsilon, b, \lambda, y) = \int_0^\infty x^{\upsilon-1} e^{-(\lambda b)\sqrt{x}} \left\{ \sum_{k=0}^{\infty} \frac{(\lambda)_k}{k!} e^{-(bk)\sqrt{x}} \right\} \begin{matrix} \sin \\ \cos \end{matrix} (xy)dx. \tag{59}$$

Upon the use of binomial expansion (52) in the above eq. (41), then we get after evaluation (37). The results (38) to (40) are derived from (29) to (32) by putting $\Theta(k) = (\lambda)_k$ and $c = b$.

Corollary 1.4.5. *Analytical expressions of the Ramanujan's integrals $\boldsymbol{R}_{S,\,C}(m, n)$ holds true for non-negative integer m and positive rational number n [21, 22].*

$$\boldsymbol{R}_{S,C}(m, n) = \int_0^\infty \frac{x^m}{\{-1 + \exp(2\pi\sqrt{x})\}} \begin{matrix} \sin \\ \cos \end{matrix} (\pi n x)dx,$$
$$= (n\pi)^{-m-1} \sum_{k=0}^{\infty} \sum_{\ell=0}^{\infty} \frac{1}{\ell!} \left\{ \frac{-(2\pi + 2\pi k)}{\sqrt{n\pi}} \right\}^{\ell} \Gamma\left(m + 1 + \frac{\ell}{2}\right) \begin{matrix} \sin \\ \cos \end{matrix} \left(\frac{m\pi}{2} + \frac{\ell\pi}{4}\right), \tag{60}$$

$$= (n\pi)^{-m-1} \sum_{k=0}^{\infty} \sum_{j=0}^{3} \frac{1}{j!} \left\{ \frac{-(2\pi + 2\pi k)}{\sqrt{n\pi}} \right\}^{j} \Gamma\left(m + 1 + \frac{j}{2}\right) \begin{matrix} \sin \\ \cos \end{matrix} \left(\frac{m\pi}{2} + \frac{j\pi}{4}\right) \times$$
$$\times {}_2F_3\left(\begin{matrix} \Delta\left(2; \dfrac{2m+j+2}{2}\right); \\ \Delta^*(4; 1+j); \end{matrix} \; \frac{-\pi^2}{4n^2} \left\{ \frac{(2)_k}{(1)_k} \right\}^4 \right), \tag{61}$$

$$= \frac{m! \begin{matrix} \sin \\ \cos \end{matrix} \left(\frac{m\pi}{2}\right)}{(n\pi)^{m+1}} \sum_{k=0}^{\infty} {}_2F_3\left(\begin{matrix} \dfrac{m+1}{2}, \dfrac{m+2}{2}; \\ \dfrac{1}{4}, \dfrac{1}{2}, \dfrac{3}{4}; \end{matrix} \; -\frac{\pi^2}{4n^2} \left\{ \frac{(2)_k}{(1)_k} \right\}^4 \right)$$
$$- \frac{\left(\frac{3}{2}\right)_m \begin{matrix} \sin \\ \cos \end{matrix} \left(\frac{m\pi}{2} + \frac{\pi}{4}\right)}{(\pi)^m (n)^{m+\frac{3}{2}}} \sum_{k=0}^{\infty} \left\{ \frac{(2)_k}{(1)_k} \right\} {}_2F_3\left(\begin{matrix} \dfrac{2m+3}{4}, \dfrac{2m+5}{4}; \\ \dfrac{1}{2}, \dfrac{3}{4}, \dfrac{5}{4}; \end{matrix} \; \frac{-\pi^2}{4n^2} \left\{ \frac{(2)_k}{(1)_k} \right\}^4 \right)$$
$$- \frac{(2)(m+1)! \begin{matrix} \sin \\ \cos \end{matrix} \left(\frac{m\pi}{2}\right)}{(\pi)^m (n)^{m+2}} \sum_{k=0}^{\infty} \left\{ \frac{(2)_k}{(1)_k} \right\}^2 {}_2F_3\left(\begin{matrix} \dfrac{m+2}{2}, \dfrac{m+3}{2}; \\ \dfrac{3}{4}, \dfrac{5}{4}, \dfrac{3}{2}; \end{matrix} \; \frac{-\pi^2}{4n^2} \left\{ \frac{(2)_k}{(1)_k} \right\}^4 \right)$$
$$+ \frac{\left(\frac{5}{2}\right)_m \begin{matrix} \sin \\ \cos \end{matrix} \left(\frac{m\pi}{2} + \frac{\pi}{4}\right)}{(\pi)^{m-1} (n)^{m+\frac{5}{2}}} \sum_{k=0}^{\infty} \left\{ \frac{(2)_k}{(1)_k} \right\}^3 {}_2F_3\left(\begin{matrix} \dfrac{2m+5}{4}, \dfrac{2m+7}{4}; \\ \dfrac{5}{4}, \dfrac{3}{2}, \dfrac{7}{4}; \end{matrix} \; \frac{-\pi^2}{4n^2} \left\{ \frac{(2)_k}{(1)_k} \right\}^4 \right), \tag{62}$$

Proof: The results (42) to (44) are obtained from (37), (38), (39) and (40) by putting $\upsilon = m + 1, b = 2\pi$, $\lambda = 1$ and $y = n\pi$.

5. Closed form infinite summation formulas

We have derived some closed forms of infinite summation formulas involving hypergeometric functions ${}_0F_1$, ${}_1F_2$, and ${}_2F_3$ [21, 22].

$$\sum_{k=0}^{\infty}\left[{}_1F_2\left(\begin{matrix}1\;; \\ \frac{1}{4},\ \frac{3}{4};\end{matrix}\quad \frac{-\pi^2}{4}\left\{\frac{(2)_k}{(1)_k}\right\}^4\right)\right]\frac{\pi}{\sqrt{2}}\sum_{k=0}^{\infty}\left[\left\{\frac{(2)_k}{(1)_k}\right\}{}_0F_1\left(\begin{matrix}-; \\ \frac{1}{2};\end{matrix}\quad \frac{-\pi^2}{4}\left\{\frac{(2)_k}{(1)_k}\right\}^4\right)\right]$$
$$+\frac{\pi^2}{\sqrt{2}}\sum_{k=0}^{\infty}\left[\left\{\frac{(2)_k}{(1)_k}\right\}^3{}_0F_1\left(\begin{matrix}-; \\ \frac{3}{2};\end{matrix}\quad \frac{-\pi^2}{4}\left\{\frac{(2)_k}{(1)_k}\right\}^4\right)\right] = \frac{\pi\sqrt{2}-4}{8}, \tag{63}$$

$$\sum_{k=0}^{\infty}\left[{}_1F_2\left(\begin{matrix}1\;; \\ \frac{1}{4},\ \frac{3}{4};\end{matrix}\quad \frac{-\pi^2}{16}\left\{\frac{(2)_k}{(1)_k}\right\}^4\right)\right]-\frac{\pi}{2}\sum_{k=0}^{\infty}\left[\left\{\frac{(2)_k}{(1)_k}\right\}{}_0F_1\left(\begin{matrix}-; \\ \frac{1}{2};\end{matrix}\quad \frac{-\pi^2}{16}\left\{\frac{(2)_k}{(1)_k}\right\}^4\right)\right]$$
$$+\frac{\pi^2}{4}\sum_{k=0}^{\infty}\left[\left\{\frac{(2)_k}{(1)_k}\right\}^3{}_0F_1\left(\begin{matrix}-; \\ \frac{3}{2};\end{matrix}\quad \frac{-\pi^2}{16}\left\{\frac{(2)_k}{(1)_k}\right\}^4\right)\right] = \frac{\pi-2}{8}, \tag{64}$$

$$\sum_{k=0}^{\infty}\left[{}_1F_2\left(\begin{matrix}1\;; \\ \frac{1}{4},\ \frac{3}{4};\end{matrix}\quad -\pi^2\left\{\frac{(2)_k}{(1)_k}\right\}^4\right)\right]-\pi\sum_{k=0}^{\infty}\left[\left\{\frac{(2)_k}{(1)_k}\right\}{}_0F_1\left(\begin{matrix}-; \\ \frac{1}{2};\end{matrix}\quad \pi^2\left\{\frac{(2)_k}{(1)_k}\right\}^4\right)\right]$$
$$+2\pi^2\sum_{k=0}^{\infty}\left[\left\{\frac{(2)_k}{(1)_k}\right\}^3{}_0F_1\left(\begin{matrix}-; \\ \frac{3}{2};\end{matrix}\quad -\pi^2\left\{\frac{(2)_k}{(1)_k}\right\}^4\right)\right] = \frac{\pi-3}{8}. \tag{65}$$

$$\sum_{k=0}^{\infty}\left[{}_2F_3\left(\begin{matrix}1,\ \frac{3}{2}\;; \\ \frac{1}{4},\ \frac{1}{2},\ \frac{3}{4};\end{matrix}\quad -\pi^2\left\{\frac{(2)_k}{(1)_k}\right\}^4\right)\right]$$
$$-\frac{3\pi}{2}\sum_{k=0}^{\infty}\left[\left\{\frac{(2)_k}{(1)_k}\right\}{}_1F_2\left(\begin{matrix}\frac{7}{4}\;; \\ \frac{1}{2},\ \frac{3}{4};\end{matrix}\quad -\pi^2\left\{\frac{(2)_k}{(1)_k}\right\}^4\right)\right]$$
$$+5\pi^2\sum_{k=0}^{\infty}\left[\left\{\frac{(2)_k}{(1)_k}\right\}^3{}_1F_2\left(\begin{matrix}\frac{9}{4}\;; \\ \frac{5}{4},\ \frac{3}{2};\end{matrix}\quad -\pi^2\left\{\frac{(2)_k}{(1)_k}\right\}^4\right)\right] = \frac{1}{32}(4\pi-13), \tag{66}$$

$$\sum_{k=0}^{\infty}\left[{}_2F_3\left(\begin{matrix}1, \frac{3}{2} ; \\ \frac{1}{4}, \frac{1}{2}, \frac{3}{4};\end{matrix} \quad \frac{-\pi^2}{16}\left\{\frac{(2)_k}{(1)_k}\right\}^4\right)\right]$$
$$-\frac{3\pi}{4}\sum_{k=0}^{\infty}\left[\left\{\frac{(2)_k}{(1)_k}\right\}{}_1F_2\left(\begin{matrix}\frac{7}{4} ; \\ \frac{1}{2}, \frac{3}{4};\end{matrix} \quad \frac{-\pi^2}{16}\left\{\frac{(2)_k}{(1)_k}\right\}^4\right)\right]$$
$$+\frac{5\pi^2}{8}\sum_{k=0}^{\infty}\left[\left\{\frac{(2)_k}{(1)_k}\right\}^3{}_1F_2\left(\begin{matrix}\frac{9}{4} ; \\ \frac{5}{4}, \frac{3}{2};\end{matrix} \quad \frac{-\pi^2}{16}\left\{\frac{(2)_k}{(1)_k}\right\}^4\right)\right] = \frac{\pi^2}{16}\left(\frac{3}{\pi}-\frac{1}{2}-\frac{5}{\pi^2}\right), \tag{67}$$

$$\sum_{k=0}^{\infty}\left[\left\{\frac{(2)_k}{(1)_k}\right\}{}_2F_3\left(\begin{matrix}\frac{7}{4}, \frac{9}{4} ; \\ \frac{1}{2}, \frac{3}{4}, \frac{5}{4};\end{matrix} \quad \frac{-\pi^2}{16}\left\{\frac{(2)_k}{(1)_k}\right\}^4\right)\right]$$
$$-\frac{16}{5}\sum_{k=0}^{\infty}\left[\left\{\frac{(2)_k}{(1)_k}\right\}^2{}_2F_3\left(\begin{matrix}2, \frac{5}{2} ; \\ \frac{3}{4}, \frac{5}{4}, \frac{3}{2};\end{matrix} \quad \frac{-\pi^2}{16}\left\{\frac{(2)_k}{(1)_k}\right\}^4\right)\right]$$
$$+\frac{7\pi}{6}\sum_{k=0}^{\infty}\left[\left\{\frac{(2)_k}{(1)_k}\right\}^3{}_2F_3\left(\begin{matrix}\frac{9}{4}, \frac{11}{4} ; \\ \frac{5}{4}, \frac{3}{2}, \frac{7}{4};\end{matrix} \quad \frac{-\pi^2}{16}\left\{\frac{(2)_k}{(1)_k}\right\}^4\right)\right] = \frac{\pi^2}{60}\left(\frac{5}{\pi}-\frac{5}{\pi^2}-1\right), \tag{68}$$

$$\sum_{k=0}^{\infty}\left[\left\{\frac{(2)_k}{(1)_k}\right\}{}_0F_1\left(\begin{matrix}-; \\ \frac{1}{2};\end{matrix} \quad -\frac{\pi^2}{4}\left\{\frac{(2)_k}{(1)_k}\right\}^4\right)\right]$$
$$-2\sqrt{2}\sum_{k=0}^{\infty}\left[\left\{\frac{(2)_k}{(1)_k}\right\}^2{}_1F_2\left(\begin{matrix}1; \\ \frac{3}{4}, \frac{5}{4};\end{matrix} \quad \frac{-\pi^2}{4}\left\{\frac{(2)_k}{(1)_k}\right\}^4\right)\right]$$
$$+\pi\sum_{k=0}^{\infty}\left[\left\{\frac{(2)_k}{(1)_k}\right\}^3{}_0F_1\left(\begin{matrix}-; \\ \frac{3}{2};\end{matrix} \quad \frac{-\pi^2}{4}\left\{\frac{(2)_k}{(1)_k}\right\}^4\right)\right] = \frac{\sqrt{2}-1}{4}, \tag{69}$$

$$\sum_{k=0}^{\infty}\left[\left\{\frac{(2)_k}{(1)_k}\right\}{}_0F_1\left(\begin{matrix}-; \\ \frac{1}{2};\end{matrix} \quad \frac{-\pi^2}{16}\left\{\frac{(2)_k}{(1)_k}\right\}^4\right)\right]$$
$$-2\sum_{k=0}^{\infty}\left[\left\{\frac{(2)_k}{(1)_k}\right\}^2{}_1F_2\left(\begin{matrix}1; \\ \frac{3}{4}, \frac{5}{4};\end{matrix} \quad \frac{-\pi^2}{16}\left\{\frac{(2)_k}{(1)_k}\right\}^4\right)\right] \tag{70}$$
$$+\frac{\pi}{2}\sum_{k=0}^{\infty}\left[\left\{\frac{(2)_k}{(1)_k}\right\}^3{}_0F_1\left(\begin{matrix}-; \\ \frac{3}{2};\end{matrix} \quad \frac{-\pi^2}{16}\left\{\frac{(2)_k}{(1)_k}\right\}^4\right)\right] = \frac{1}{4},$$

$$\sum_{k=0}^{\infty}\left[\left\{\frac{(2)_k}{(1)_k}\right\}{}_0F_1\left(\begin{matrix}-; \\ \frac{1}{2};\end{matrix}\quad \frac{-\pi^2}{64}\left\{\frac{(2)_k}{(1)_k}\right\}^4\right)\right]$$
$$-\sqrt{2}\sum_{k=0}^{\infty}\left[\left\{\frac{(2)_k}{(1)_k}\right\}^2{}_1F_2\left(\begin{matrix}1; \\ \frac{3}{4},\ \frac{5}{4};\end{matrix}\quad \frac{-\pi^2}{64}\left\{\frac{(2)_k}{(1)_k}\right\}^4\right)\right]$$
$$+\frac{\pi}{4}\sum_{k=0}^{\infty}\left[\left\{\frac{(2)_k}{(1)_k}\right\}^3{}_0F_1\left(\begin{matrix}-; \\ \frac{3}{2};\end{matrix}\quad \frac{-\pi^2}{64}\left\{\frac{(2)_k}{(1)_k}\right\}^4\right)\right]=\frac{3\sqrt{2}-2}{4}, \tag{71}$$

$$\sum_{k=0}^{\infty}\left[\left\{\frac{(2)_k}{(1)_k}\right\}{}_0F_1\left(\begin{matrix}-; \\ \frac{1}{2};\end{matrix}\quad \frac{-\pi^2}{144}\left\{\frac{(2)_k}{(1)_k}\right\}^4\right)\right]$$
$$-\frac{2\sqrt{3}}{3}\sum_{k=0}^{\infty}\left[\left\{\frac{(2)_k}{(1)_k}\right\}^2{}_1F_2\left(\begin{matrix}1; \\ \frac{3}{4},\ \frac{5}{4};\end{matrix}\quad \frac{-\pi^2}{144}\left\{\frac{(2)_k}{(1)_k}\right\}^4\right)\right]$$
$$+\frac{\pi}{6}\sum_{k=0}^{\infty}\left[\left\{\frac{(2)_k}{(1)_k}\right\}^3{}_0F_1\left(\begin{matrix}-; \\ \frac{3}{2};\end{matrix}\quad \frac{-\pi^2}{144}\left\{\frac{(2)_k}{(1)_k}\right\}^4\right)\right]=\frac{13\sqrt{3}-12}{12}, \tag{72}$$

$$\sum_{k=0}^{\infty}\left[\left\{\frac{(2)_k}{(1)_k}\right\}{}_0F_1\left(\begin{matrix}-; \\ \frac{1}{2};\end{matrix}\quad -\pi^2\left\{\frac{(2)_k}{(1)_k}\right\}^4\right)\right]$$
$$-4\sum_{k=0}^{\infty}\left[\left\{\frac{(2)_k}{(1)_k}\right\}^2{}_1F_2\left(\begin{matrix}1; \\ \frac{3}{4},\ \frac{5}{4};\end{matrix}\quad -\pi^2\left\{\frac{(2)_k}{(1)_k}\right\}^4\right)\right]$$
$$+2\pi\sum_{k=0}^{\infty}\left[\left\{\frac{(2)_k}{(1)_k}\right\}^3{}_0F_1\left(\begin{matrix}-; \\ \frac{3}{2};\end{matrix}\quad -\pi^2\left\{\frac{(2)_k}{(1)_k}\right\}^4\right)\right]=\frac{1}{8\pi}, \tag{73}$$

$$\sum_{k=0}^{\infty}\left[\left\{\frac{(2)_k}{(1)_k}\right\}{}_0F_1\left(\begin{matrix}-; \\ \frac{1}{2};\end{matrix}\quad \frac{-25\pi^2}{16}\left\{\frac{(2)_k}{(1)_k}\right\}^4\right)\right]$$
$$-2\sqrt{5}\sum_{k=0}^{\infty}\left[\left\{\frac{(2)_k}{(1)_k}\right\}^2{}_1F_2\left(\begin{matrix}1; \\ \frac{3}{4},\ \frac{5}{4};\end{matrix}\quad \frac{-25\pi^2}{16}\left\{\frac{(2)_k}{(1)_k}\right\}^4\right)\right]$$
$$+\frac{5\pi}{2}\sum_{k=0}^{\infty}\left[\left\{\frac{(2)_k}{(1)_k}\right\}^3{}_0F_1\left(\begin{matrix}-; \\ \frac{3}{2};\end{matrix}\quad \frac{-25\pi^2}{16}\left\{\frac{(2)_k}{(1)_k}\right\}^4\right)\right]=\frac{8\sqrt{5}-15}{100}. \tag{74}$$

Proof: When $m = 0$ with $n = 1,2,\frac{1}{2}$ in the eqs. (42) to (44) and comparing with the eqs. (17), (18), and (19), we get the results (46), (47) and (48) respectively. In view of the hypergeometric functions (70), (71) and (72), we can express the above results

(46) to (48) in terms of cosine, sine and Lommel functions. Our results (46) to (48) are convergent in view of the convergence condition of ${}_pF_q(\cdot)$ series, when $p \leq q$, and for all $|z| < \infty$.

Similarly, we derive (49) to (51) by putting $m = 1, n = \frac{1}{2}$; $m = 1, n = 2$ and $m = 2, n = 2$ in the eqs. (42) and (44) and finally comparing with (3), (4) and (5). When $m = 0$ with $n = 1,2,4,6,\frac{1}{2},\frac{2}{5}$ in the eqs. (42) and (44) and comparing with (10) to (15), we get the rest of results (52) to (57) respectively. In view of the hypergeometric functions (53), (54) and (55), we can express the above results (52) to (57) in terms of cosine, sine and Lommel functions. Our results (49) to (57) are convergent in view of the convergence condition of ${}_pF_q(\cdot)$ series, when $p \leq q$, and for all $|z| < \infty$.

6. Concluding remarks

We have derived analytical expressions of the infinite Fourier sine and cosine transforms related to Ramanujan's integrals as an infinite sum of ordinary hypergeometric functions ${}_2F_3$, with suitable convergence conditions. Moreover, as applications of Ramanujan's integrals $\mathbf{R}_S(m, n)$, some closed form infinite summation formulas associated with hypergeometric functions ${}_1F_2$, ${}_2F_3$ and ${}_0F_1$ are evaluated. It is hoped that other such integrals can also be evaluated in a similar way. We conclude by remarking that various new results and applications can be obtained from our general theorem by appropriate choice of the parameters υ,λ,b,c,y and fixed sequence $\{\Theta(k)\}_{k=0}^{\infty}$ in $I_C^*(\upsilon, b, c, \lambda, y)$.

Acknowledgements

One of the authors (S.A. Dar) acknowledges financial support from the University Grants Commission of India for the award of a Dr. D.S. Kothari Post Doctoral Fellowship(DSKPDF) (Grant number F.4-2/2006 (BSR)/MA/20-21/0061).

Author details

Showkat Ahmad Dar* and M. Kamarujjama
Department of Applied Mathematics, Aligarh Muslim University, Aligarh, India

*Address all correspondence to: showkatjmi34@gmail.com

References

[1] Carslaw HS. Introduction to the Theory of Fourier's Series and Integrals. Martin's street, London: Macmillan and co., limited st; 1921

[2] Erdélyi A, Magnus W, Oberhettinger F, Tricomi FG. Higher Transcendental Functions. Vol. 1. New York, Toronto and London: McGraw-Hill; 1953

[3] Erdélyi A, Magnus W, Oberhettinger F, Tricomi FG. Tables of Integral Transforms. Vol. 1. New York, Toronto and London: McGraw-Hill; 1954

[4] Gradshteyn IS, Ryzhik IM. Table of Integrals, Series, and Products. 8th ed. USA: Academic Press is an imprint of Elsevier; 2015

[5] Oberhettinger F. Tables of Fourier Transforms and Fourier Transforms of Distributions. Berlin: Springer Verlag; 1990

[6] Sneddon NI. Fourier Transforms. Newyork: McGraw Hill Book Company, Inc; 1951

[7] Ramanujan S. Some definite integrals connected with Gauss's sums. Messenger of Mathematics;**XLIV**(2015): 75-86

[8] Oberhettinger F. Tables of Bessel Transforms. Berlin, Heidelberg, New York: Springer-Verlag; 1972

[9] Rainville ED. Special Functions. New York: Macmillan Company; 1960 Reprinted by Chelsea Publishing Company, Bronx, New York. 1971

[10] Srivastava HM, Manocha HL. A Treatise on Generating Functions. New York, Chichester, Brisbane and Toronto: Halsted Press (Ellis Horwood Limited, Chichester, U.K.), John Wiley and Sons; 1984

[11] Srivastava HM. A note on certain identities involving generalized hypergeometric series. Nederl. Akad. Wetensch. Proc. Ser. A 82=Indag. Math. 1979;**41**:191-201

[12] Kilbas AA, Saigo M. H-Transforms: Theory and Applications (Analytical Methods and Special Functions). Vol. 9. Boca Raton, London, New York, Washington, D.C.: CRC Press Company; 2004

[13] Kilbas AA, Saigo M, Trujillo JJ. On the generalized Wright function. Fractional Calculus and Applied Analysis. 2002;**5**(4):437-460

[14] Andrews GE, Berndt BC. Ramanujan's Lost Notebook. New York: Part IV. Springer-Verlag; 2013

[15] Berndt BC, Straub A. Certain integrals arising from Ramanujan's notebooks, symmetry, Integrability and geometry: Methods and applications. SIGMA. 2015;**11**(83):1-11

[16] Berndt BC. Integrals associated with Ramanujan and elliptic functions. The Ramanujan Journal. 2016;**41**:369-389

[17] Ramanujan S. Collected Papers. Cambridge: Cambridge University Press; 1927 reprinted by Chelsea, New York, 1962, reprinted by Amer. Math. Soc., Providence, RI, 2000

[18] Ramanujan S. The Lost Notebook and Other Unpublished Papers. New Delhi: Narosa; 1988

[19] Meyer JL. A generalization of an integral of Ramanujan. The Ramanujan Journal. 2007;**14**:79-88

[20] Qureshi MI, Khan IH. Ramanujan integrals and other definite integrals

associated with Gaussian hypergeometric functions. South East Asian Journal of Mathematics and Mathematical Sciences. 2005;**4**(1):39-52

[21] Qureshi MI, Dar SA. Generalizations of Ramanujan's integral associated with infinite Fourier cosine transforms in terms of hypergeometric functions and its applications. Kyungpook National University. 2020;**60**:781-795

[22] Qureshi MI, Dar SA. Generalizations of Ramanujan's integral associated with infinite Fourier sine transforms in terms of hypergeometric functions and its applications. Montes Taurus Journal of Pure and Applied Mathematics. 2021; **3**(3):216-226

Chapter 5

Perspective Chapter: Cascaded-Resonator-Based Recursive Harmonic Analysis

Miodrag D. Kušljević

Abstract

It is well known that recursive algorithms for harmonic analysis have better characteristics in terms of monitoring the change of the spectrum in comparison to methods based on the processing of blocks of consecutive samples, such as, for example, discrete Fourier transform (DFT). This property is particularly important when applying spectral estimation in real-time systems. One of the recursive algorithms is the resonator-based one. The approach of the parallel cascades of multiple resonators (MR) with the common feedback has been generalized as the cascaded-resonator (CR)-based structure for recursive harmonic analysis. The resulting filters of the CR structure can be finite impulse response (FIR) type or the infinite impulse response (IIR) ones as a computationally more efficient solution, optimizing the frequency responses of all harmonics simultaneously. In the case of the IIR filter, the unit characteristic polynomial present in the FIR filter is replaced with an optimized characteristic polynomial of the transfer function. Such a change does not lead to an increase in computing requirements and changes only the resonator gain values. By using a conveniently linearized iterative algorithm for stability control purpose, based on the Rouche's theorem, the iterative linear-programming-based or the constrained linear least-squares (CLLS) optimization techniques can be used.

Keywords: cascaded-resonator (CR)-based filter, constrained linear least squares (CLLS), discrete Fourier transformation (DFT), Taylor-Fourier transformation (TFT), harmonic analysis, IIR filter, linear programming (LP), multiple-resonator (MR)-based filter

1. Introduction

In recent years, a lot of various algorithms for harmonic analysis have been proposed in the literature. Good surveys of some techniques are presented in Refs. [1, 2]. The discrete Fourier transform (DFT)-based method, as a mainstream approach, is widely used for harmonic analysis, thanks to its low computational burden, especially with the fast Fourier transform (FFT). However, errors arise when the power system is operating at off-nominal frequency, especially under dynamic conditions. Harmonic estimates under oscillating conditions were recently proposed in several studies. A huge volume of papers has been written on harmonics tracking in power systems. The focus of recent literature has been on preprocessing and postprocessing methods for fixed-sample-rate algorithms surrounding a core DFT (or similar) analysis with a fixed number of samples [3, 4].

Idea of considering a dynamic model to better estimate the fundamental and harmonic phasors has been emerging in Refs. [4–9], and its importance has been pointed out in Ref. [10]. In Ref. [9], the discrete Taylor-Fourier transform (TFT) was proposed as an extension of the full DFT. The TFT by using a dynamic model of the signal extends and improves estimations obtained by DFT [9, 11]. This transformation corresponds to an FIR filter bank with a maximally flat frequency response. Each filter in the bank has maximum flat gain around the harmonic frequency and near-ideal attenuation around the other harmonics. This results in less distortion of the signal and less influence of disturbances present in the signal. In this way, the periodicity restriction assumed by the Fourier analysis is mitigated. As result, so obtained reconstruction is more accurate than the reconstruction obtained through DFT. When harmonics are narrow-band pass signals with spectral density confined into the flat-gain harmonic intervals, the coefficients of the TFT provide good estimates of the first derivatives of their complex envelopes. The digital TFT formulation in a matrix form that facilitates its implementation with the FFT to reduce the computational complexity of its straightforward implementation has been given in [11].

The multiple-resonator (MR)-based recursive estimators have been introduced in Ref. [12]. In Ref. [13], the MR-based observer structure is proposed for the implementation of TFT. Their good properties are provided by their parallel form, a recursive implementation, and good sensitivity properties assured by the infinite loop gain at the resonator frequencies [14]. Multiple zeros also provide reinforcing of the required attenuations and zero-gain flatness at the harmonic components with a high overall attenuation in the stopbands. For the known frequency of the periodic signal, the estimator based on resonators with common feedback enables the estimation of Fourier components even in cases when the sampling rate is not synchronized with the signal frequency. Also, this harmonic analyzer shows robustness in real conditions where there is noise and nonlinearity of the analog part of the equipment. MR-based harmonic analysis provides better performances of the spectral estimation than the single-resonator-based observer that corresponds to the classical DFT estimator.

This approach has been generalized as the cascaded-resonator (CR)-based structure for harmonic analysis. In Ref. [15], the cascaded-dispersed-resonator-based (CDR-based) structure for harmonic analysis is proposed. Although the design objectives in Refs. [15, 16] are different, the design technique is the same in both cases. In Ref. [16], the task is to replace multiple resonators with a cascade of single resonators. In this way, for the design purpose, it is possible to use the classic Lagrange interpolation technique instead of the more complex Hermitian interpolation. The condition that the poles are distributed in a narrow band around the resonant frequencies, as close as possible to each other, which is however limited by numerical accuracy. In Ref. [15], the task is to arrange the poles in the cascade in such a way as to enable optimal attenuation in the entire range around the harmonic frequencies. Practically, the only difference is in the arrangement of the resonator poles around the harmonic frequencies. The frequency deviation issue can be resolved by adaptive estimators based on the actual frequency feedback. This approach has drawbacks as a stability issue, due to an internal delay. Instead of that, usage of the external module for the fundamental frequency estimation is proposed in Ref. [17].

2. Cascaded-resonator-based structure harmonic analysis

Figure 1 shows the block diagram of the K-type CR-based harmonic analyzer. The structure includes $(K+1)(2M+2)$ resonators with poles $\{z_{m,k}, m = -M, \dots, 0, \dots, M+1, k = 0, 2, \dots K\}$, placed in the $2M+2$ cascades each of them having $K+1$ cascaded complex poles on the unit circle around related harmonic frequency [13, 15, 16]. Each resonator has its belonging complex gains $g_{m,k}$. A complete set of resonator cascades is connected in parallel in a common feedback loop. The gains of the transfer functions at the resonator frequencies are equal to unity due to the infinite loop gain at these frequencies. The number of cascades (for the coherent sampling) is $2M+2$ (M is a number of harmonics) and depends on ω_1, because the condition $M\omega_1 < \pi$ has to be satisfied. The $\omega_1 = 2\pi f_1/f_S$, f_1 and f_S are the nominal fundamental frequency and sampling rate. The overall system order is $(K+1)(2M+2)$.

We have in every mth channel of the structure, as an internal transfer function

$$H_m(z) = \frac{V_m^F(z)}{E(z)} = z^{-1}\sum_{k=0}^{K} \frac{g'_{m,k}}{\prod_{i=0}^{k-1}\left(1 - z_{m,i}z^{-1}\right)}, \quad g'_{m,k} = \prod_{i=k}^{K} g_{m,i} \tag{1}$$

where $m = -M, \dots, 0, \dots, M+1$, $k = 0, 1, \dots, K$, and $\prod_{i=0}^{k-1}\left(1 - z_{m,i}z^{-1}\right) = 1$ for $k = 0$. $V_m^F(z)$ is the total feedback signal corresponding to mth channel, composed as the linear combination of the output of each resonator, i.e., each channel contributes to the filter output with $K+1$ complex weights $\left\{g_{m,k}, k = 0, 1, \dots, K\right\}$.

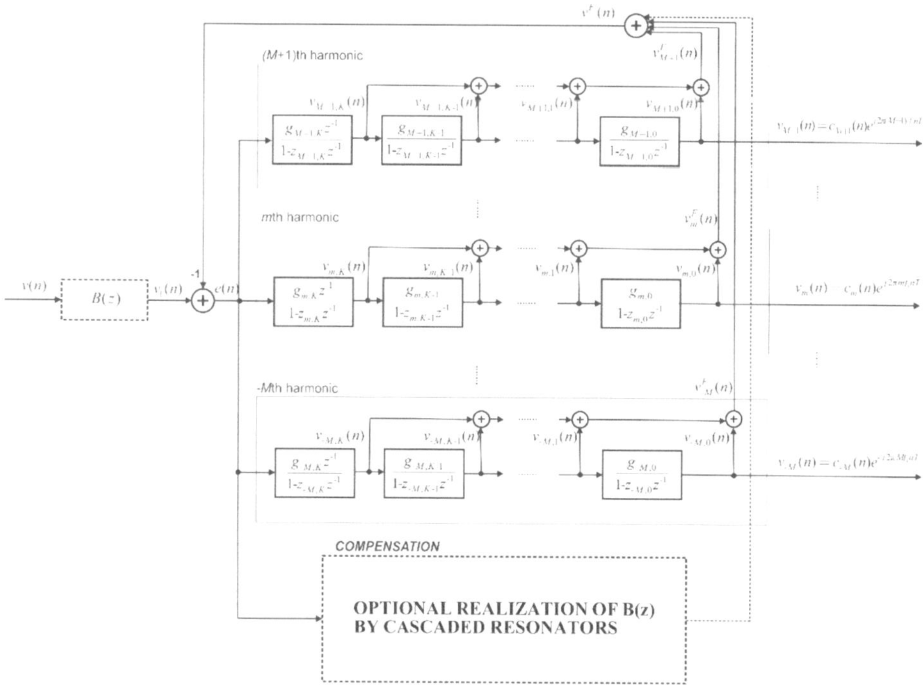

Figure 1.
Block diagram of the K-type CR-based harmonic analyzer.

The closed transfer function for every channel m has the form of [15, 16].

$$T_{m,0}(z) = \frac{V_{m,0}(z)}{V_1(z)} = g'_{m,0}\frac{z^{-1}P_m(z)}{A(z)}, \quad P_m(z) = \prod_{\substack{n=-M \\ n \neq m}}^{M+1} \prod_{i=0}^{K} (1 - z_{n,i}z^{-1}), \tag{2}$$

$$T_{m,k}(z) = \frac{V_{m,k}(z)}{V_1(z)} = g'_{m,k}\frac{z^{-1}P_m(z)\prod_{i=0}^{k-1}(1 - z_{m,i}z^{-1})}{A(z)} \tag{3}$$

$$A(z) = \prod_{n=-M}^{M+1} \prod_{i=0}^{K} (1 - z_{n,i}z^{-1}) + z^{-1} \sum_{n=-M}^{M+1} \left[P_n(z) \sum_{k=0}^{K} \left(g'_{n,k} \prod_{i=0}^{k-1} (1 - z_{n,i}z^{-1}) \right) \right] \tag{4}$$

It can be seen that all poles of the resonators are mapped to the zeros of the transfer function $T_{m,0}(z)$ due to the common feedback, with the exception of the poles belonging to the cascade of the harmonic m, which are automatically canceled by the poles that generated them. In differentiators transfer functions $\{T_{m,k}(z), k = 1, \dots, K\}$, $(k \neq 0)$, zeros $\{z_{m,i}, i = 0, \dots, k - 1\}$, originated from poles in mth channel, exist providing zero gain.

From Eq. (2), it is obvious that the filter corresponding to mth-channel provides the maximally flatness property in the stop band around the remaining harmonic frequencies. For small pole displacements, Δf frequency response reshaping is negligible in comparison to the multiple-resonator case [16]. It is important to mention that the lower border of Δf is limited by the computational accuracy. On the other hand, avoiding of the multiple poles allows design by the direct usage of the classical Lagrange interpolation formula rather than the Hermite one.

Although the characteristic polynomial of the transfer functions can be chosen in different ways, under some conditions it is possible to choose one so that the error is driven to zero in exactly $(K + 1)(2M + 2)$ samples. This is provided by what is called a dead-beat observer, for which the coefficients are calculated from the condition that the observer has deadbeat settling, i.e., it finds the unknown state within at most $(K + 1)(2M + 2)$ steps. That leads to FIR filters (with $A(z) = 1$) in each channel. This way, although the structure is realized by resonators, which are IIR filters, the resulting filters in each channel are FIR type. **Figure 2** shows frequency responses of $T_{1,0}(z) = g'_{1,0}z^{-1}P_1(z)$ (corresponding to the fundamental component) of the dead-beat observer for the first up to the sixth order of resonator multiplicity $(K = 0, 1, \dots, 5)$. It is observed that (quasi) MR structures with a higher order multiplicity of poles provide smaller sidelobes and thus ensure a lower sensitivity to noise and to harmonic and interharmonic disturbances. The negative effect is the increase in the order of the filter, which increases the group delay and response time of the filter, as well as the numerical complexity. Due to this feature, large values of the resonator multiplicity could be inconvenient in the control application. The case $K = 0$ corresponds to a classic DFT estimator, while the cases $K > 0$ correspond to the TFT.

Frequency responses of zeroth-, first-, and second-order differentiator discrete FIR filters corresponding to the transfer functions $T_{m,k}(z)$ related to the fundamental component $(m = 1, k = 0, 1, 2)$, for $K = 2$ and $K = 3$ (the third- and forth-order resonator structure) are given in **Figure 3**.

In order to obtain wider flatness intervals in the pass band, the feedback signals $V_m^F(z)$ could be used for harmonic estimation instead of $V_{m,0}(z)$ [13]. The global

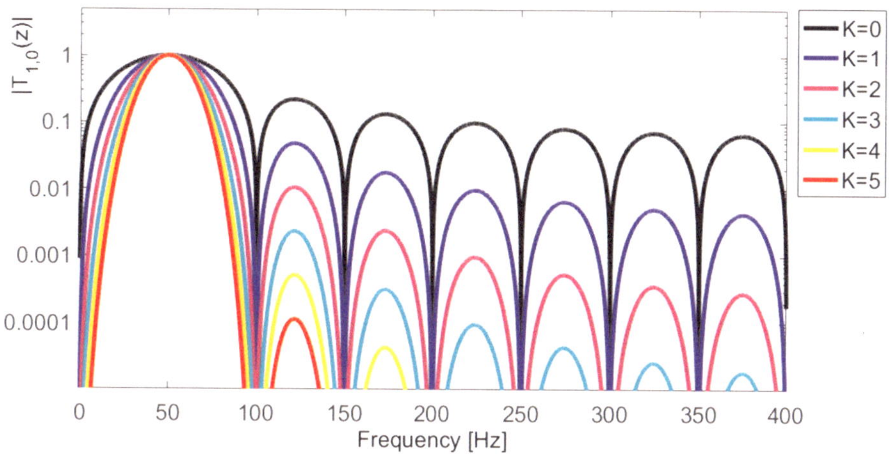

Figure 2.
Frequency responses for $T_{1,0}(z)$ (the zeroth differentiator of the first harmonic) for $K = 0, 1, \ldots, 5$ (the first- to sixth-order resonator structure).

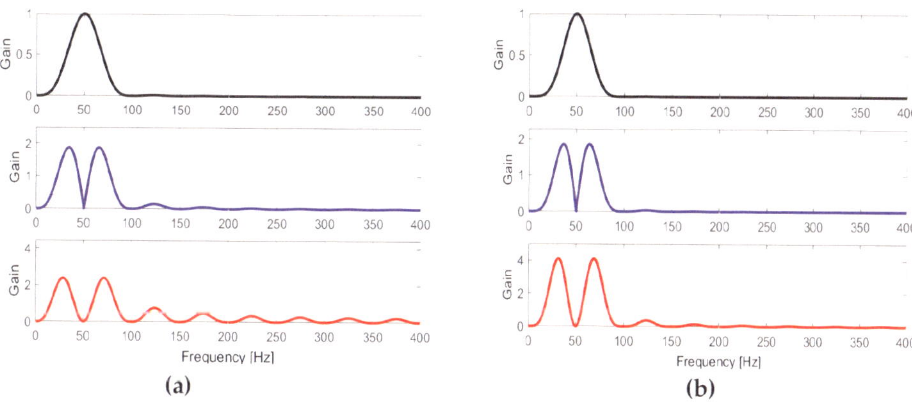

Figure 3.
Frequency response of the zeroth-, first-, and second-order discrete FIR filters corresponding to $T_{m,k}(z)$, ($m = 1$ and $k = 0, 1, 2$), related to the fundamental component, for $f_s = 800\text{Hz}$ and $f_1 = 50\text{Hz}$, for (a) $K = 2$ estimator (the third-order resonator structure) and (b) $K = 3$ estimator (the fourth-order resonator structure).

transfer function of the feedback loop is a sum of the transfer functions of all $K + 1$ differentiators $T_m^F(z) = V_m^F(z)/V(z) = \sum_{k=0}^{K} T_{m,k}(z)$. The frequency responses of the estimation of the fundamental component obtained by $V_{m,0}(z)$ and estimation obtained by $V_m^F(z)$ are given together in **Figure 4a**. The good properties of the filters corresponding to the transfer functions $T_m^F(z)$ are related to the phase responses. Frequency responses have a zero phase response in the frequency bands around the harmonic frequencies, which means that in those frequencies the group delay is equal to zero. Bad properties are high resonant gains at the edges of bandwidths and high sidelobes. The zero flat gains in the stop band are preserved, although their intervals are narrowed. It should be mentioned that the peaks of the interharmonic gains and the side lobes increase by the multiplicity of the resonators (**Figure 4b**).

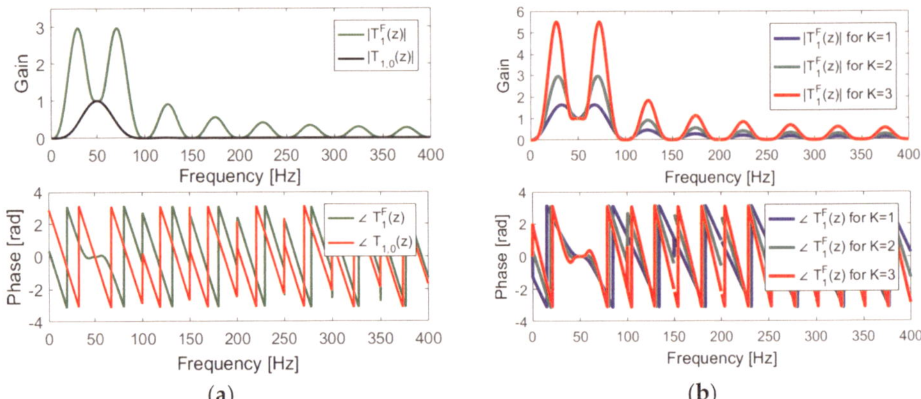

Figure 4.
Frequency responses of (a) $T_{1,0}(z)$ and $T_1^F(z)$ for $K = 2$ estimator (the third-order cascade) and (b) $T_1^F(z)$ for $K = 1$, $K = 2$ and $K = 3$, related to the fundamental component, for $f_s = 800$Hz and $f_1 = 50$Hz.

2.1 Optimization problem statement

In order to adapt the achieved digital differentiators to their ideal frequency responses around the harmonic frequencies, it is possible to modify the filters transfer functions. An optimization technique is utilized to reshape frequency responses of the filters transfer functions, avoiding resonant frequency peaks and reducing a group delay simultaneously, that can be rather important in control applications. The optimization task can also be different, e.g., maximization of the selectivity.

The transfer function of the extended structure, including the compensation FIR filter $B(z)$, for the mth component channel is as follows:

$$T_{m,0}^{AB}(z) = \frac{V_{m,0}(z)}{V_1(z)} = B(z)T_{m,0}(z) = \left[g'_{m,0}z^{-1}P_m(z)\right]\frac{B(z)}{A(z)} = \left[g'_{m,0}z^{-1}P_m(z)\right]\frac{\boldsymbol{q}_B\boldsymbol{x}_B}{1+\boldsymbol{q}_A\boldsymbol{x}_A} \tag{5}$$

where

$$B(z) = b_0 + b_1z^{-1} + \cdots + b_{N_B-1}z^{-(N_B-1)} + b_{N_B}z^{-N_B},$$
$$A(z) = 1 + a_1z^{-1} + \cdots + a_{N_A-1}z^{-(N_A-1)} + a_{N_A}z^{-N_A},$$

$$\boldsymbol{x}_B = [b_0\ b_1 \cdots\ \ b_{N_B-1}\ \ b_{N_B}]^{\mathrm{T}},\ \boldsymbol{x}_A = [a_1\ \ a_2\ \ \cdots\ \ a_{N_A-1}\ \ a_{N_A}]^{\mathrm{T}},\ \boldsymbol{x} = \left[\boldsymbol{x}_B{}^{\mathrm{T}}\ \ \boldsymbol{x}_A{}^{\mathrm{T}}\right]^{\mathrm{T}},$$

$$\boldsymbol{q}_B = \left[1\ \ z^{-1}\ \ z^{-2}\ \ \ z^{-(N_B-1)}\ \ z^{-N_B}\right],\ \boldsymbol{q}_A = \left[z^{-1}\ \ z^{-2}\ \ \ z^{-(N_A-1)}\ \ z^{-N_A}\right].$$

The polynomial $A(z)$ does not cause any additional computation and only the polynomial $B(z)$ represents an additional numerical burden. Even more, in some cases, it is possible to choose $B(z) = b_0$ which causes only one additional multiplication. Nevertheless, for the purposes of design, we will consider the IIR filter $B(z)/A(z)$ as a common compensation for the total set of FIR filters $P_m(z)$, $m = -M, \ldots, 0, \ldots, M+1$.

With a given weighting function $W(\omega)$, the weighted Chebyshev error between the desired and actual frequency responses is defined as follows:

$$J^1(\boldsymbol{x}) = \max_{\omega \in \Omega} \quad W(\omega)\left|T^{AB}\left(e^{j\omega}\right) - H^d\left(e^{j\omega}\right)\right| \tag{6}$$

where $H^d\left(e^{j\omega}\right)$ is the desired frequency response in angular frequency ω specified in the frequency region Ω (or a union of several compact frequency bands) of the interests and $0 \leq \omega \leq \pi$.

The sum of squares of absolute values of errors in N_F angular frequencies as follows:

$$J^2(\boldsymbol{x}) = \sum_{i=1}^{N_F} W(z_i)\left[T^{AB}(z_i) - H^d(z_i)\right] \tag{7}$$

where $z_i = e^{\,j\omega_i}$.

In order to minimize the error $J^1(\boldsymbol{x})$ defined in Eq. (6), a new variable δ can be introduced and the problem reformulated as follows:

$$\begin{aligned} &\text{minimize} \quad \delta \\ &\text{subject to} \quad |E(\omega_i)| \leq \delta, \ \ \omega_i \in \Omega, \ \ i = 1, 2, \cdots, N_F \end{aligned} \tag{8}$$

where $E(z_i) = W(z_i)\left[T^{AB}(z_i) - H^d(z_i)\right], z_i = \exp\left(j\omega_i\right)$, for the total number N_F of points defined in Ω.

Further, it is:

$$\frac{W(z_i)}{|A(z_i)|}\left|T(z_i)B(z_i) - A(z_i)H^d(z_i)\right| \leq \delta \tag{9}$$

In Ref. [18], a suitable method has been described to linearize the error function $J^1(\boldsymbol{x})$ such that the design problem can be solved by the linear programming (LP) method. However, this method neglects the denominator part $|A(z_i)|$. In Ref. [19], the performance of the LP method was improved by eliminating the above drawback by using the following iterative constraints scheme:

$$\frac{W(z_i)}{\left|A^{(k-1)}(z_i)\right|}\left|T(z_i)B(z_i) - A^{(k)}(z_i)H^d(z_i)\right| \leq \delta \tag{10}$$

For the sake of notational simplicity, we denote

$$\frac{W(z_i)}{\left|A^{(k-1)}(z_i)\right|}\left|\left[T(z_i)\boldsymbol{q}_B\big|_{z=z_i}, \ \ -H^d(z_i)\boldsymbol{q}_A\big|_{z=z_i}\right]\boldsymbol{x} - H^d(z_i)\right| \leq \delta \tag{11}$$

The vector of unknown coefficients $\boldsymbol{x}$ is expanded with an additional variable δ, so that the expanded vector of unknowns is obtained:

$$\boldsymbol{x}^\delta = [\boldsymbol{x} \quad \delta]^T. \tag{12}$$

The constraints defined by inequality (11) refer to the frequency ranges in which the error optimization is performed. In addition, sometimes it is necessary to keep the error within predefined limits, such as for example the gains in the stopbands and/or the transition bands:

$$\frac{1}{\left|A^{(k-1)}(z_i)\right|}\left|\left[T(z_i)\boldsymbol{q_B}\Big|_{z=z_i}, \quad -H^d(z_i)\boldsymbol{q_A}\Big|_{z=z_i}\right]\boldsymbol{x} - H^d(z_i)\right| \le l_i \tag{13}$$

$z_i = \exp(j\omega_i), i = 1, 2, \cdots, N_G$

l_i , $i = 1, 2, \cdots, N_G$, are fixed borders of the absolute values of the set of complex error $E(z)$ along assemblies $z_i = \exp(j\omega_i), i = 1, 2, \cdots, N_G$.

If one wants to ensure unity gain in harmonic frequencies, the following condition must be met:

$$\frac{B(z)}{A(z)}\bigg|_{z=z_m} = 1, \text{ i.e. } \quad B(z_m) = A(z_m). \tag{14}$$

where $m = -M, \dots, 0, \dots, M+1$.

In a matrix form, it can be written as follows:

$$\boldsymbol{q_m x} = 1. \tag{15}$$

where $\boldsymbol{q} = \begin{bmatrix} \boldsymbol{q_B} & -\boldsymbol{q_A} \end{bmatrix}$, $\boldsymbol{q_m} = \boldsymbol{q}\big|_{z=z_m}$.

Complex equality constraints (15) can be written as follows

$$\begin{bmatrix} \text{Re}\{\boldsymbol{q_m}\} \\ \text{Im}\{\boldsymbol{q_m}\} \end{bmatrix}\boldsymbol{x} = \begin{bmatrix} 1 \\ 0 \end{bmatrix}, \quad m = -M, \dots, 0, \dots, M+1. \tag{16}$$

2.2 Linearization of constraints

The inequalities (11) and (13) are nonlinear. The convex semi-infinite programming can be applied [20], thanks to the quadratic property of the functions. Furthermore, a convenient approximation of these inequalities by the system of the linear ones [21–25] allows us to solve this constrained optimization problem through the LP or the constrained linear least-squares (CLLS) optimization technique.

It is valid:

$$|E(z_i)| = |E(z_i)|\left(\cos^2\alpha_i + \sin^2\alpha_i\right) = \text{Re}\{E(z_i)\}\cos\alpha_i + \text{Im}\{E(z_i)\}\sin\alpha_i \tag{17}$$

where $\alpha_i = \arg\{E(z_i)\}$.

Since α_i is not known a priory, the nonlinear constraints in Eq. (8) can be approximated by the system of linear constraints:

$$\text{Re}\{E(z_i)\}\cos\alpha_{i,j} + \text{Im}\{E(z_i)\}\sin\alpha_{i,j} \le \delta \tag{18}$$

where $i = 1, 2, \cdots, N_F$. If we choose L equidistantly distributed angles then it is $\alpha_{i,j} = \alpha_{i,0} + (j-1)2\pi/L$, where $j = 1, 2, \cdots, L$. **Figure 5** shows that approximations by square and octagon ($L = 4$ and $L = 8$, respectively) allow only rough approximations. A higher accuracy is obtained by increasing L. Herein, $L = 32$ is used.

Let us define as following for frequency point i:

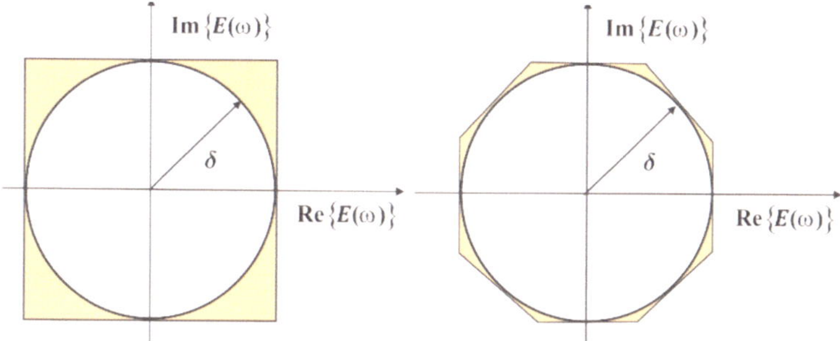

Figure 5.
Approximation of a cycle with a square and an octagon.

$$W^A(z_i) = \frac{W(z_i)}{\left|A^{(k-1)}(z_i)\right|}, \quad \boldsymbol{g}_i = \left[T(z_i)\boldsymbol{q}_B\big|_{z=z_i}, \quad -H^d(z_i)\boldsymbol{q}_A\big|_{z=z_i}\right]. \tag{19}$$

Hence, Eq. (13) can be linearized and written in a matrix form

$$W^A(z_i)\boldsymbol{A}_i'\boldsymbol{x} \le W^A(z_i)\boldsymbol{b}_i' + l_i\mathbf{1}_{L\times 1}, \quad i = 1, 2, \cdots, N_G \tag{20}$$

where matrix $\boldsymbol{A}_i'$ and vector $\boldsymbol{b}_i'$ are given by

$$\boldsymbol{A}_i' = \begin{bmatrix} \text{Re}\{\boldsymbol{g}_i\}\cos\alpha_{i1} + \text{Im}\{\boldsymbol{g}_i\}\sin\alpha_{i1} \\ \text{Re}\{\boldsymbol{g}_i\}\cos\alpha_{i2} + \text{Im}\{\boldsymbol{g}_i\}\sin\alpha_{i2} \\ \vdots \\ \text{Re}\{\boldsymbol{g}_i\}\cos\alpha_{iL} + \text{Im}\{\boldsymbol{g}_i\}\sin\alpha_{iL} \end{bmatrix}, \ \boldsymbol{b}_i' = \begin{bmatrix} \text{Re}\{H^d(z_i)\}\cos\alpha_{i1} + \text{Im}\{H^d(z_i)\}\sin\alpha_{i1} \\ \text{Re}\{H^d(z_i)\}\cos\alpha_{i2} + \text{Im}\{H^d(z_i)\}\sin\alpha_{i2} \\ \vdots \\ \text{Re}\{H^d(z_i)\}\cos\alpha_{iL} + \text{Im}\{H^d(z_i)\}\sin\alpha_{iL} \end{bmatrix}.$$

and l_i is a constraint limit of the error in the point z_i.

Using matrix notation, and collecting inequality linearization systems in all settled frequency points, (20) becomes the following linear form:

$$\boldsymbol{A}'\boldsymbol{x} \le \boldsymbol{b}' \quad \text{or} \quad \left[\boldsymbol{A}' \quad \mathbf{0}_{(N_G L)\times 1}\right]\boldsymbol{x}^\delta \le \boldsymbol{b}' \tag{21}$$

where matrix $\boldsymbol{A}'$ and vector $\boldsymbol{b}'$ are given by

$$\boldsymbol{A}' = \begin{bmatrix} W^A(z_1)\boldsymbol{A}_1' \\ W^A(z_2)\boldsymbol{A}_2' \\ \vdots \\ W^A(z_{N_G})\boldsymbol{A}_{N_G}' \end{bmatrix}, \ \boldsymbol{b}' = \begin{bmatrix} W^A(z_1)\boldsymbol{b}_1' + l_1\mathbf{1}_{L\times 1} \\ W^A(z_2)\boldsymbol{b}_2' + l_2\mathbf{1}_{L\times 1} \\ \vdots \\ W^A(z_{N_G})\boldsymbol{b}_{N_G}' + l_{N_G}\mathbf{1}_{L\times 1} \end{bmatrix}.$$

In case of the (11), we have:

$$\left[W^A(z_i)\boldsymbol{A}_i' \quad -\mathbf{1}_{L\times 1}\right]\boldsymbol{x}^\delta \le W^A(z_i)\boldsymbol{b}_i', \quad i = 1, 2, \cdots, N_F \tag{22}$$

or in a matrix notation

$$\left[\boldsymbol{A}' \quad -\mathbf{1}_{(N_F L)\times 1}\right]\boldsymbol{x}^\delta \le \boldsymbol{b}' \tag{23}$$

where

$$A' = \begin{bmatrix} W^A(z_1)A'_1 \\ W^A(z_2)A'_2 \\ \vdots \\ W^A(z_{N_F})A'_{N_F} \end{bmatrix}, \ b' = \begin{bmatrix} W^A(z_1)b'_1 \\ W^A(z_2)b'_2 \\ \vdots \\ W^A(z_{N_F})b'_{N_F} \end{bmatrix}.$$

2.3 Design (optimization) approach 1: CLLS minimization

An objective is to find a minimum of the sum of squares of absolute values of $\boldsymbol{hx} - d$ in the assembly of the N_F selected frequencies subject to the vector $\boldsymbol{x}$

$$\min_x \sum_{i=1}^{N_F} |\boldsymbol{h}_i\boldsymbol{x} - d_i|^2 \tag{24}$$

where $\boldsymbol{h}_i = W(z_i)\boldsymbol{g}_i/\left|A^{(k-1)}(z_i)\right|$ and $d_i = W(z_i)H^d(z_i)/\left|A^{(k-1)}(z_i)\right|$.
If we apply the following equality

$$|\boldsymbol{h}_i\boldsymbol{x} - d_i|^2 = \text{Re}^2\{\boldsymbol{h}_i\boldsymbol{x} - d_i\} + \text{Im}^2\{\boldsymbol{h}_i\boldsymbol{x} - d_i\} = \|\boldsymbol{C}_i\boldsymbol{x} - \boldsymbol{d}_i\|_2^2 \tag{25}$$

where $\boldsymbol{C}_i = \begin{bmatrix} \text{Re}\{\boldsymbol{h}_i\} \\ \text{Im}\{\boldsymbol{h}_i\} \end{bmatrix}$, $\boldsymbol{d}_i = \begin{bmatrix} \text{Re}\{d_i\} \\ \text{Im}\{d_i\} \end{bmatrix}$, (24) can be written in a matrix form:

$$\min_x \|\boldsymbol{Cx} - \boldsymbol{d}\|_2^2. \tag{26}$$

where $\boldsymbol{C}$ and $\boldsymbol{d}$ include $\boldsymbol{C}_i$ and $\boldsymbol{d}_i$, respectively, $i = 1, 2, \cdots, N_F$.

The constrained linear least squares (CLLS) is an optimization problem that deals with the maximization or minimization of a linear function called the objective function subject to linear constraints. Summarizing (16), (21), and (26), the CLLS problem is formalized as follows:

$$\min_{\boldsymbol{x}} \frac{1}{2}\|\boldsymbol{Cx} - \boldsymbol{d}\|_2^2 \ \text{ subject to } \ \boldsymbol{Ax} \leq \boldsymbol{b} \ \left(\text{and} \begin{bmatrix} \text{Re}\{\boldsymbol{q}_m\} \\ \text{Im}\{\boldsymbol{q}_m\} \end{bmatrix} \boldsymbol{x} = \begin{bmatrix} 1 \\ 0 \end{bmatrix}\right), \quad (m = -M, \ldots, 0, \ldots M+1) \tag{27}$$

where $\boldsymbol{A}$ and $\boldsymbol{b}$ include $\boldsymbol{A}'$ and $\boldsymbol{b}'$, respectively, defined in (21), for all frequency points in which the constraints are defined.

2.4 Design (optimization) approach 2: minimax optimization

The LP optimization problem can be formalized in the following way:

$$\text{minimize} \ \ \mathbf{c}\,\boldsymbol{x}^\delta \ \ \text{subject to } \boldsymbol{Ax}^\delta \leq \boldsymbol{b} \ \left(\text{and} \begin{bmatrix} \text{Re}\{\boldsymbol{q}_m\} & 0 \\ \text{Im}\{\boldsymbol{q}_m\} & 0 \end{bmatrix} \boldsymbol{x}^\delta = \begin{bmatrix} 1 \\ 0 \end{bmatrix}\right), \quad (m = -M, \ldots, 0, \ldots M+1) \tag{28}$$

where $\mathbf{c} = \begin{bmatrix} \mathbf{0}_{1\times(N_A+N_B+1)} & 1 \end{bmatrix}$, and $\boldsymbol{A}$ and $\boldsymbol{b}$ include $\boldsymbol{A}'$ and $\boldsymbol{b}'$, respectively, for all frequency points in which the constraints or objective functions are defined in (21) and (23), respectively.

3. IIR cascaded-resonator-based harmonic analysis

In accordance with the prevailing trends in works dealing with this issue, in the initial works [13, 16, 22, 23, 25, 26] the resulting filters of CR structures were of the FIR type. Later, in [27], IIR filters were used, which represent a computationally more efficient solution [28, 29]. The unit characteristic polynomial of the transfer function is replaced by the optimized one. Such a change does not lead to an increase in the volume of numerical calculations and only requires a change in the gain values associated with the resonators. Since the optimization of frequency characteristics for all harmonics is carried out at the same time, it is possible to obtain frequency responses of the same shape. By using a linearized iterative scheme [30] based on Rouche's theorem with the aim of stability control, it is possible to use iterative optimization techniques based on LP or CLLS.

3.1 Problem statement

The task of optimization is to design a filter $B(z)/A(z)$ where the order of the characteristic polynomial $A(z)$ is $N_A = (K+1)(2M+2)$ and the polynomial $B(z)$ is of order N_B. We seek to find a causal stable rational function $T^{AB}_{m,0}(z) = \left[g'_{m,0}z^{-1}P_m(z)\right]B(z)/A(z)$ for $m = -M, \ldots, 0, \ldots, M+1$ that best approximates $H^d_m(e^{j\omega})$.

In order to make the notation as simple and short as possible, let us form a virtual transfer function so that in each bandwidth centered in mf_1 with width of f_1, i.e., for $f \in (mf_1 - f_1/2, mf_1 + f_1/2]$, it corresponds to the transfer function belonging to the harmonic m. It follows:

$$T^A(z) = \frac{T(z)}{A(z)}, \quad T(z) = g'_{m,0}z^{-1}P_m(z) \tag{29}$$

for $f \in (mf_1 - f_1/2, mf_1 + f_1/2], m = -M, \ldots, 0, \ldots, M+1$.

In addition, we define a unique transfer function

$$T^{AB}(z) = B(z)T^A(z) = B(z)\frac{T(z)}{A(z)} \tag{30}$$

Similarly, a virtual unique desired transfer function in an angular frequency ω has the following form [27]:

$$H^d\left(e^{j\omega}\right) = \begin{cases} e^{-j2\pi\tau(f-mf_1)/f_S}, & \text{for } f \in [mf_1 - f_{PB}, mf_1 + f_{PB}] \\ 0, & \text{for } f \in [mf_1 - f_1/2, mf_1 - f_{SB}] \cup [mf_1 + f_{SB}, mf_1 + f_1/2] \end{cases} \tag{31}$$

where pass and stop bands are defined by f_{PB} and f_{SB}, respectively. A desired group delay in the passband is denoted as τ.

3.2 Stability constraint

To obtain a stable IIR filter $T(z)$, stability constraint must be imposed on the coefficient vector $\boldsymbol{x}_A$. In [30], the more convenient stability condition which is based on Rouche's theorem was proposed.

Rouche's Theorem. *If $f(z)$ and $g(z)$ are analytic inside and on a closed contour C, and $|g(z)| < |f(z)|$ on C, then $f(z)$ and $g(z)+f(z)$ have the same number of zeros inside C.*

Let

$$f(z) = z^{N_A}A(z) = z^{N_A} + a_1 z^{N_A-1} + \cdots + a_{N_A-1}z + a_{N_A}, \tag{32}$$

$$g(z) = z^{N_A}\Delta(z) = \delta_0 z^{N_A} + \delta_1 z^{N_A-1} + \cdots + \delta_{N_A-1}z + \delta_{N_A}, \tag{33}$$

where $\Delta(z)$ is the update of the characteristic polynomial $A(z)$ of the transfer function at each iteration step. Since the functions $f(z)$ and $g(z)$ are analytic, except at $z = \infty$, and have the same zeros as $A(z)$ and $\Delta(z)$, according to Rouche's theorem, if the polynomial $A^{(k-1)}(z)$ in the iteration step, $k-1$ has all its zeros inside a circle of radius ρ $(0<\rho<1)$ with the center at the origin of the complex plane, then also the polynomial in the iteration step k given by [30]

$$A^{(k)}(z) = A^{(k-1)}(z) + \alpha\Delta^{(k)}(z), \;\; 0<\alpha<1 \tag{34}$$

will retain the zeros within this circle provided that in step k the following condition satisfied

$$\left|\Delta^{(k)}(z)\right| \le \left|A^{(k-1)}(z)\right|, \;\; |z| = \rho. \tag{35}$$

If (34) is included in (35), we get

$$\left|\bar{A}^{(k)}(z) - \bar{A}^{(k-1)}(z)\right| \le \alpha\left|A^{(k-1)}(z)\right| \tag{36}$$

where $\bar{A}^{(k)}(z) = A^{(k)}(z) - 1$, $\bar{A}^{(k-1)}(z) = A^{(k-1)}(z) - 1$, or in a matrix notation:

$$\left|\left[\mathbf{0}_{1\times(N_B+1)}, \;\; \boldsymbol{q}_A\big|_{z=z_i}\right]\boldsymbol{x} - \bar{A}^{(k-1)}(z_i)\right| \le \alpha\left|A^{(k-1)}(z_i)\right| \tag{37}$$

As for the initial value of the vector $\boldsymbol{x}$, it is simplest to take $\boldsymbol{x}^{(0)} = \mathbf{0}$, when all the roots of the polynomial $A^{(0)}(z)$ lie within the circle of radius ρ $(0<\rho<1)$.

If constraint (37) is applied to a sufficiently dense set of points lying on a circle of radius ρ $(|z| = \rho)$, of total length N_S, we get

$$\boldsymbol{A}^{\mathrm{S}}\boldsymbol{x} \le \boldsymbol{b}^{\mathrm{S}} \tag{38}$$

where matrix $\boldsymbol{A}_i^{\mathrm{S}}$ and vector $\boldsymbol{b}_i^{\mathrm{S}}$ are given by

$$\boldsymbol{A}_i^S = \begin{bmatrix} \mathbf{0}_{1\times(N_B+1)}, \mathrm{Re}\left\{\boldsymbol{q}_A\big|_{z=z_i}\right\}\cos\alpha_{i1} + \mathrm{Im}\left\{\boldsymbol{q}_A\big|_{z=z_i}\right\}\sin\alpha_{i1} \\ \mathbf{0}_{1\times(N_B+1)}, \mathrm{Re}\left\{\boldsymbol{q}_A\big|_{z=z_i}\right\}\cos\alpha_{i2} + \mathrm{Im}\left\{\boldsymbol{q}_A\big|_{z=z_i}\right\}\sin\alpha_{i2} \\ \vdots \\ \mathbf{0}_{1\times(N_B+1)}, \mathrm{Re}\left\{\boldsymbol{q}_A\big|_{z=z_i}\right\}\cos\alpha_{iL} + \mathrm{Im}\left\{\boldsymbol{q}_A\big|_{z=z_i}\right\}\sin\alpha_{iL} \end{bmatrix},$$

$$\boldsymbol{b}_i^S = \begin{bmatrix} \mathrm{Re}\left\{\bar{A}^{(k-1)}(z_i)\right\}\cos\alpha_{i1} + \mathrm{Im}\left\{\bar{A}^{(k-1)}(z_i)\right\}\sin\alpha_{i1} + \alpha\left|A^{(k-1)}(z_i)\right| \\ \mathrm{Re}\left\{\bar{A}^{(k-1)}(z_i)\right\}\cos\alpha_{i2} + \mathrm{Im}\left\{\bar{A}^{(k-1)}(z_i)\right\}\sin\alpha_{i2} + \alpha\left|A^{(k-1)}(z_i)\right| \\ \vdots \\ \mathrm{Re}\left\{\bar{A}^{(k-1)}(z_i)\right\}\cos\alpha_{iL} + \mathrm{Im}\left\{\bar{A}^{(k-1)}(z_i)\right\}\sin\alpha_{iL} + \alpha\left|A^{(k-1)}(z_i)\right| \end{bmatrix}.$$

Thus, the set of constraints (38) is added to the set of the above constraint conditions. In this way, the iterative methods mentioned above solve the LP or CLLS problem by taking into account constraints (38) in each iteration step. A fixed step size α can be used, while a gradual decrease (e.g., exponential) can help the convergence of the solution.

3.3 Resonators' gains calculation

After the polynomial $A(z)$ having been determined, the direct usage of the Lagrange interpolation formula provides the closed-form formulas [26]. It should be taken into account that these formulas are valid only in the case of single resonators. If a quasi-MR-based analyzer is designed, the resonator poles connected to the same harmonic should be arranged close enough to each other with a minimum distance that is limited by numerical precision. Its lower border depends on the resonator multiplicity and the sampling rate. A chosen displacement of 0.1 Hz allows a fair approximation for sampling frequencies up to 6.4 kHz ($M = 63$ for $f_1 = 50$ Hz) and $K = 5$ [16].

A generalized closed-form formula for gains calculation for any K for previously chosen polynomial $A(z)$ is given as follows:

$$g'_{m,k} = \left.\frac{A(z) - z^{-1}P_m(z)\sum_{j=0}^{k-1}\left[g'_{m,j}\prod_{i=0}^{j-1}\left(1 - z_{m,i}z^{-1}\right)\right]}{z^{-1}P_m(z)\prod_{i=0}^{k-1}\left(1 \quad z_{m,i}z^{-1}\right)}\right|_{z=z_{m,k}} \tag{39}$$

As a final result, the designed resonator gains are as follows:

$$g_{m,k} = g'_{m,k}/g'_{m,k+1}, \left(g'_{m,K+1} = 1\right),\ m = -M, \dots, 0, \dots, M+1, k = 0, 1, \dots, K.$$

It should be mentioned that polynomial $B(z)$ can be conveniently implemented by adding its roots as poles in additional parallel channels to the basic structure (see **Figure 1**). In this case, the existing formulas for gains calculation are valid only for the identical structure of the extension cascades (they have to consist $K + 1$ resonators). Otherwise, the formulas are not valid and need a completely new derivation.

3.4 Design example

In the next section, three demonstration examples, with frequency responses and pole-zeros maps, of the designed $K = 2$ type CR-based harmonic analyzer are shown, for $f_S = 800$ Hz and $f_1 = 50$ Hz. For a clear readability, lower values of $f_S = 800$ Hz and $M = 7$ are selected. The following parameters are prescribed in all three examples:

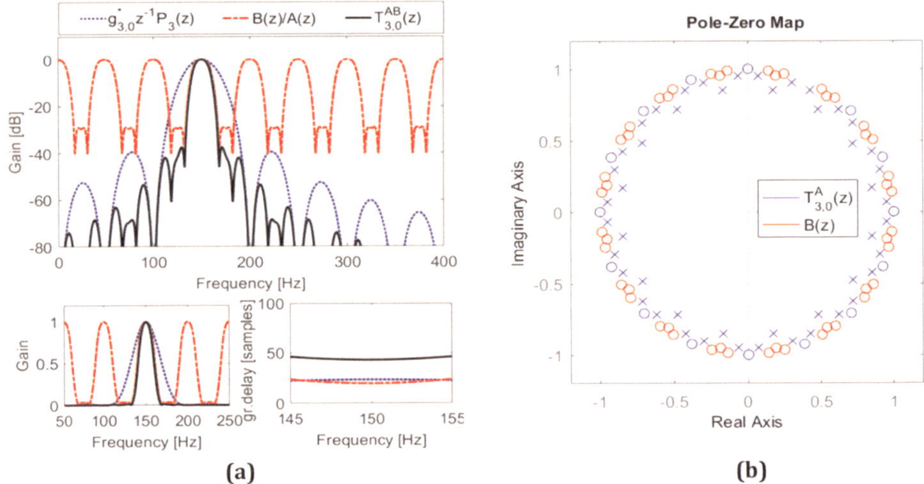

Figure 6.
(a) Frequency responses of $g'_{3,0}z^{-1}P_3(z)$, $B(z)/A(z)$ and $T^{AB}_{3,0}(z)$and (b) Pole-zero map of $T_{3,0}(z)$ and $B(z)$ (for the third harmonic), for $N_A = (K+1)(2M+2)$, $N_B = (K+1)(2M+2)$, and $\rho = 0.95$, $f_{PB} = 1.7$ Hz and τ=0.9$(K+1)(2M+2)$.

$f_{SB} = 17$ Hz, $l_{TB} = 1.005$, $W(z) = 1, \alpha_0 = 0.5$, $\rho = 0.95$, $N_A = (K+1)(2M+2)$. $\boldsymbol{x}^{(0)} = \mathbf{0}$. Other parameters were variated, depending on the chosen optimization criteria. It should mention that a large variety of optimization scenarios is possible, allowing design spectrum analyzers for a wide scope of different applications.

3.4.1 Example 1: flat-top passbands

The filters with a wider flatness in the pass band allow better signal tracking in the dynamic conditions. Since it is difficult task to provide the tracking of the parameters changes together with good attenuation in the stopband, a relatively high order of $N_B = (K+1)(2M+2)$ is settled. The desired group delay (in samples) in the passband is τ=0.9$(K+1)(2M+2)$. $f_{PB} = 1.7$ Hz. $l^{PB} = 0.01$. Obtained frequency responses (**Figure 6**) show that the passband flatness is not derogated, while the selectivity and attenuation in the stopbands are increased thanks to the zeros of the polynomial $B(z)$ which are located between the existing multiple zeros of the resonator structure that had been obtained through the common feedback. A cost is an increased total group delay which causes higher latency.

3.4.2 Example 2: narrow selective passbands

In this example, the requests for passband and transition bands are omitted, which decrease a numerical burden. To obtain high selectivity, $N_B = (K+1)(2M+2)$ is kept. The obtained frequency responses (see **Figure 7**) show that selectivity and attenuation in the stopbands are increased. This is achieved thanks to the poles of the transfer function located on a circle of radius 0.95 ($|z| = 0.95$) very close to the zeros located in the harmonic frequencies, as well as the additional zeros of the polynomial $B(z)$. Such high selectivity caused a large increase in a group delay.

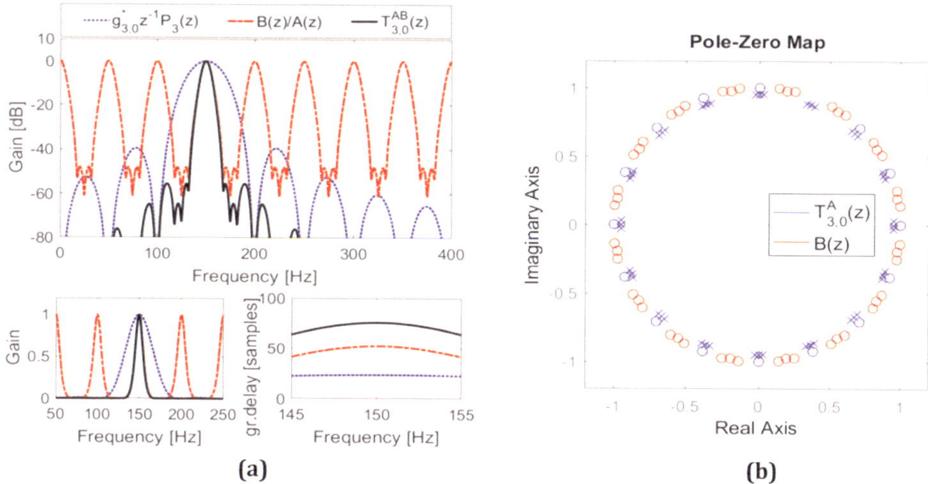

Figure 7.
(a) Frequency responses of $g'_{3,0}z^{-1}P_3(z)$, $B(z)/A(z)$ *and* $T^{AB}_{3,0}(z)$*and (b) Pole-zero map of* $T_{3,0}(z)$ *and* $B(z)$ *(for the third harmonic), for* $N_A = (K+1)(2M+2)$, $N_B = (K+1)(2M+2)$, *and* $\rho = 0.95, f_{PB} = 0$ Hz.

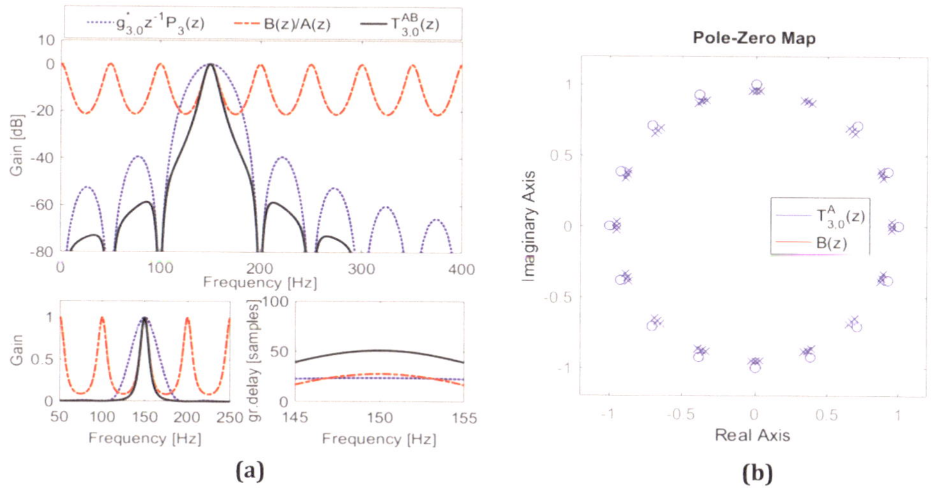

Figure 8.
(a) Frequency responses of $g'_{3,0}z^{-1}P_3(z)$, $B(z)/A(z)$ *and* $T^{AB}_{3,0}(z)$*and (b) Pole-zero map of* $T_{3,0}(z)$ *and* $B(z)$ *(for the third harmonic), for* $N_A = (K+1)(2M+2)$, $N_B = 0$, *and* $\rho = 0.95, f_{PB} = 0$ Hz.

3.4.3 Example 3: numerically cost-effective solution

This example is very similar to the previous one with different that now is $N_B = 0$, which means that there is no extension to the existing resonator structure with common feedback. Obtained frequency responses (see **Figure 8**) show that selectivity and attenuation in the stopbands are smaller than in the previous case, however, with a smaller group delay too.

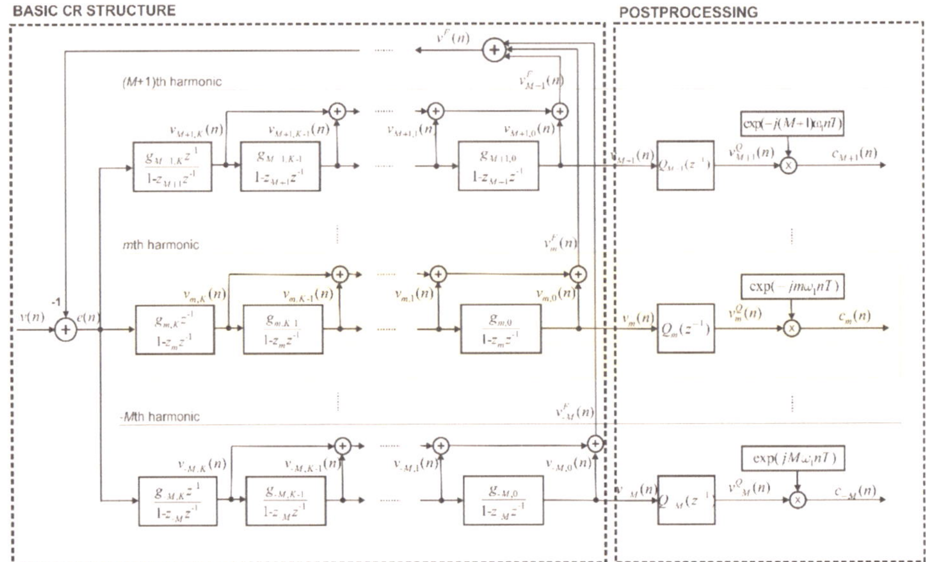

Figure 9.
Block diagram of the K-type CR-based harmonic analyzer with postprocessing.

4. FIR cascaded-resonator-based harmonic phasor estimation

Instead of the common simultaneous compensation of the frequency responses for all harmonics through the compensating filter $B(z)$ placed in the front of the parallel resonator structure, a more flexible solution is shown in **Figure 9** with postprocessing by the set of compensators $Q_m(z)$ designed particularly for each harmonic m. In this case, $A(z) = 1$ is chosen to allow the use of linear optimization techniques such as LP and CLLS.

In order to obtain an algorithm that can be utilized in a wide range of signal dynamics in a unified way and improve the frequency response, a linear combination of the differentiators' outputs in the cascade can be used [22, 25, 26]. The goal of this compromised solution was to propose a tracking-mode harmonic estimation technique. In Ref. [31], it is shown that this estimation technique exhibiting maximally flat frequency responses can be efficiently used for implementation of P-Class Compliant PMU in accordance with IEC/IEEE Standard 60255-118-1:2018 for harmonic phasors estimation. In this approach, the order of the resulted compensation filter was low and equals to the pole multiplicity. In Refs. [21, 23, 24], the proposed approach was generalized to any necessary order through the postprocessing compensation FIR filters applied to the output signals obtained by the CR structure. The drawback of this approach is that we have to use as many postprocessing FIR filters as there are harmonic phasors that we need to estimate (one estimator per one harmonic phasor). On the other hand, the advantage is that it is possible to obtain a filter bank, surrounding the core CR structure, with a set of different compensation filters corresponding to different signal dynamics.

The transfer function for every mth channel has the form of

$$T^{Q}_{m,0}(z) = \frac{V^{Q}_{m,0}(z)}{V(z)} = T_{m,0}(z)Q_m(z) = \left[g'_{m,0}z^{-1}P_m(z)\right]\boldsymbol{q}_Q\boldsymbol{x}_Q. \tag{40}$$

where $Q_m(z) = q_{m,0} + q_{m,1}z^{-1} + \cdots + q_{m,N_Q-1}z^{-(N_Q-1)} + b_{m,N_Q}z^{-N_Q}$.
$\boldsymbol{q_Q} = \begin{bmatrix} 1 & z^{-1} & z^{-2} & \ldots & z^{-(N_Q-1)} & z^{-N_Q} \end{bmatrix}$, $\boldsymbol{x_Q} = \begin{bmatrix} q_0 & q_1 & \cdots & q_{N_Q-1} & q_{N_Q} \end{bmatrix}^{\mathrm{T}}$.

Eq (40) has the same form as Eq. (5) with the following constraints: $A(z) = 1$, $N_A = 0$, $\boldsymbol{x_A} = []$, $\boldsymbol{q_A} = []$, $\boldsymbol{x} = \boldsymbol{x_Q}$, $\boldsymbol{q} = \boldsymbol{q_Q} \cdot \boldsymbol{q_B}$ is replaced by $\boldsymbol{q_Q}$, and $\boldsymbol{x_B}$ by $\boldsymbol{x_Q}$. Since $N_A = 0$, the stability constraints are not present. The desired frequency response is related only to the actual harmonic m and does not consider the frequency responses of the other ones.

4.1 Total vector gradient (TVG) calculation

The response time and delay of the estimator are directly correlated with the group delay (GD) of the filter. Due to the more complex calculation of GD, it is possible to use the gradient of the transfer function $dT^Q_{m,0}(z)/dz$, which is called the total vector gradient (TVG) here. In a flat range with small amplitude changes, TVG and GD are proportional, and optimization of one leads to optimization of the other. The first derivative of the transfer function $T^Q_{m,0}(z)$ is as follows:

$$\begin{aligned} dT^Q_{m,0}(z)/dz &= g'_{m,0}\left[-z^{-2}Q_m(z)P_m(z) + z^{-1}P_m(z)dQ_m(z)/dz + z^{-1}Q_m(z)dP_m(z)/dz\right] \\ &= g'_{m,0}z^{-1}[P_m(z)dQ_m(z)/dz + \Psi_m(z)Q_m(z)] \end{aligned} \tag{41}$$

where

$$\frac{dQ_m(z)}{dz} = -q_{m,1}z^{-2} - 2q_{m,2}z^{-3} - \cdots - N_Q q_{m,N_Q}z^{-(N_Q+1)}$$

$$\Psi_m(z) = -z^{-1}P_m(z) + dP_m(z)/dz$$

$$\frac{dP_m(z)}{dz} = P_m(z)\sum_{\substack{i=-M \\ i\neq m}}^{M+1}\sum_{k=0}^{K}\frac{z_{i,k}z^{-2}}{1 - z_{i,k}z^{-1}}.$$

Eq. (41) can be written in a matrix form as follows:

$$dT^Q_{m,0}(z)/dz = \boldsymbol{\psi_m x_{Q,m}}. \tag{42}$$

where

$$\boldsymbol{\psi_m} = \begin{bmatrix} \psi_{m,0}(z) & \psi_{m,1}(z) & \cdots & \psi_{m,N_Q-1}(z) & \psi_{m,N_Q}(z) \end{bmatrix},$$
$$\psi_{m,n}(z) = g'_{m,0}z^{-(n+1)}\left[\Psi_m\left(z^{-1}\right) - nz^{-1}P_m\left(z^{-1}\right)\right], \; n = 0, 1, \ldots, N_Q.$$

4.2 Optimization criteria

Selections of the object and functions and constraints can be very different depending on the optimization criteria scenario. Herein will be considered three criteria summarized in **Table 1** [21, 24]. In the first Criterion 1, the cost function in which absolute values are minimized is the transfer function $T^Q_{m,0}(z)$ in the stop bands.

Criteria	Object function	Desired values	Frequency range	Constrained functions	Reference values	Frequency range
Criterion 1	$T^{Q}_{m,0}(z)$	0	Stopbands	$T^{Q}_{m,0}(z)$	$T^{d}_{m}(z) = e^{-j\tau(\omega-\omega_m)}$	Passband
				$T^{Q}_{m,0}(z)$	0	Transition band
Criterion 2	$T^{Q}_{m,0}(z)$	0	Stopbands	$dT^{Q}_{m,0}(z)/dz$	0	Harmonic frequency
				$T^{Q}_{m,0}(z)$	0	Transition band
Criterion 3	$dT^{Q}_{m,0}(z)/dz$	0	Harmonic frequency	$T^{Q}_{m,0}(z)$	0	Stopband
				$T^{Q}_{m,0}(z)$	0	Transition band

Table 1.
Considered design criteria.

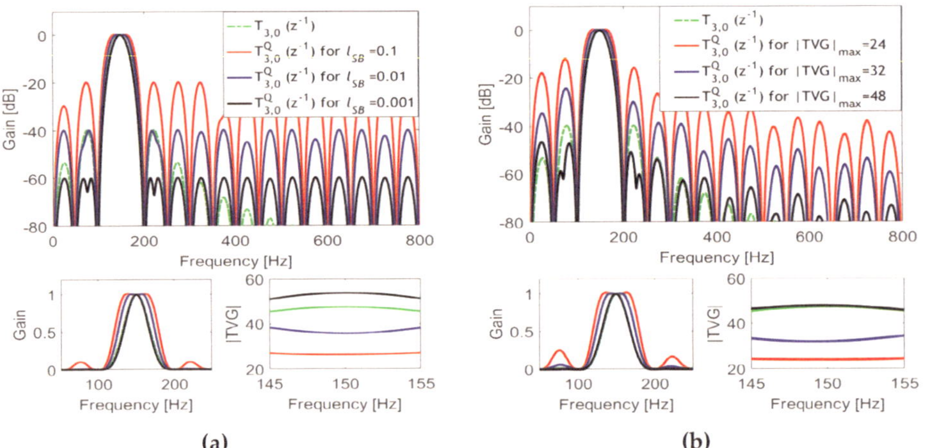

Figure 10.
Frequency responses for the basic ($T_{3,0}(z)$) *and reshaped* ($T^{Q}_{3,0}(z)$) *transfer function for* $K = 2$, *for* $f_s = 1.6$ kHz, $N_Q = 16$ *and (a)* $l_3^{SB} \in \{0.1, 0.01, 0.001\}$ *and (b)* $|TVG|_{max} \in \{24, 32, 48\}$.

Weighting function $W_m(\omega)$ is settled to 1. The function which absolute values are kept under settled limits is an error in the passband $E_m(z_i) = T^{Q}_{m,0}(z)\Big|_{z=z_i} - T^{d}_{m}(z_i)$ where $T^{d}_{m}(z_i) = \exp\left(-j\tau(\omega_i - \omega_m)\right)$. In addition to this, a control of overshoots in the transition bands is performed.

Criterion 2 is similar with Criterion 1 with the difference that the absolute value of the TVG in the harmonic frequency is limited, that is $\left|dT^{Q}_{m,0}(z)/dz\right|_{z=z_m} \leq |TVG|_{max}$, where $|TVG|_{max}$ is a maximally allowed absolute total vector gradient. Like in Criterion 1, the limitation of overshoots in the transition bands is necessary.

Criterion 3 minimizes the absolute value of the TVG in the harmonic frequency subject to the limitation of the gain in the stopband. Similarly with the previous cases, the limitation of overshoots in the transition bands is necessary.

4.3 Design example

In order to illustrate the described algorithms, examples overtaken from [21] are shown for Criteria 2 and 3 defined in **Table 1**. **Figures 10** and **11** show the frequency responses of the transfer function of the third harmonic $T_{3,0}^{Q}(z)$ in the case of $K = 2$. For Criterion 3, the maximum allowed gains in the stopband was selected as $l_m^{SB} \in \{0.1, 0.01, 0.001\}$, which corresponds to attenuations of $\{20, 40, 60\}$ dB. For Criterion 2, the maximum value of $|TVG|$ in the harmonic frequencies z_m, $|TVG|_{max} \in \{24, 32, 48\}$ was selected. Zoomed amplitude and $|TVG|$ characteristics around the harmonic frequency are shown in the inset figures at the bottom of the figures. It can be seen that the higher value of N_Q gives a smaller value of $|TVG|$ and wider bandwidth. It is also visible that for smaller values $|TVG|$ sidelobes are larger,

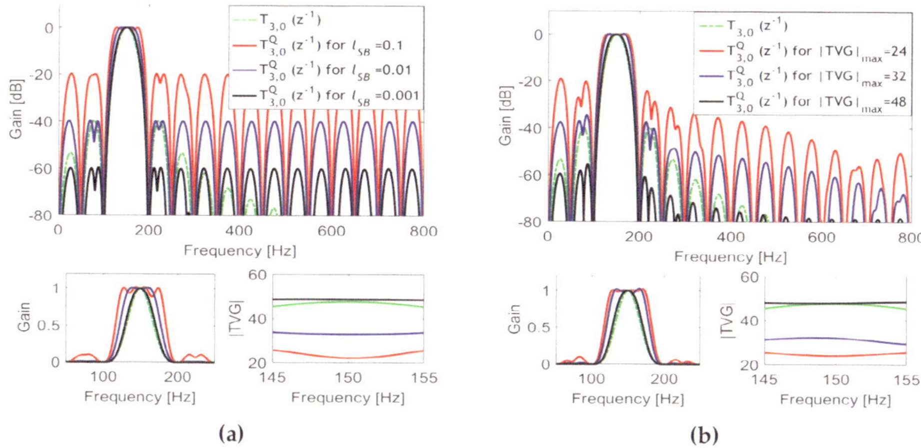

Figure 11.
Frequency responses for the basic ($T_{3,0}(z)$) and reshaped ($T_{3,0}^{Q}(z)$) transfer function for $K = 2$, for $f_s = 1,6$ kHz, $N_Q = 32$ and (a) $l_3^{SB} \in \{0.1, 0.01, 0.001\}$ and (b) $|TVG|_{max} \in \{24, 32, 48\}$.

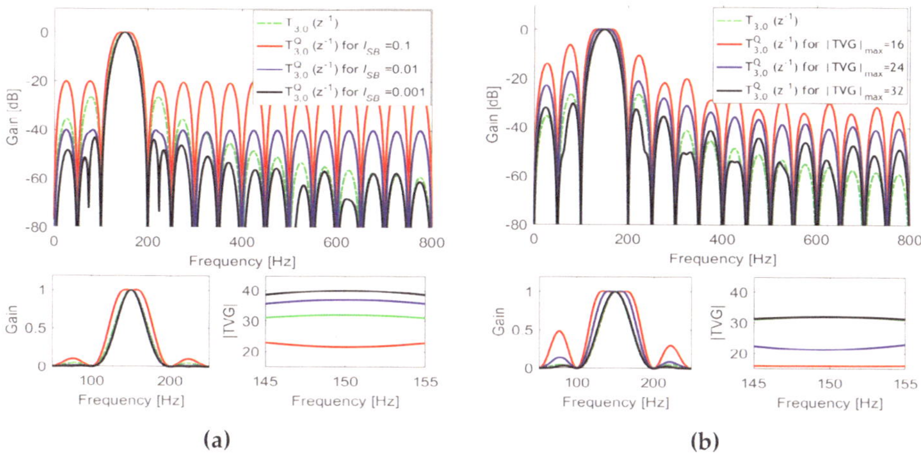

Figure 12.
Frequency responses for the basic ($T_{3,0}(z)$) and reshaped ($T_{3,0}^{Q}(z)$) transfer function for $K = 1$, for $f_s = 1,6$ kHz, $N_Q = 16$ and (a) $l_3^{SB} \in \{0.1, 0.01, 0.001\}$ and (b) $|TVG|_{max} \in \{16, 24, 32\}$.

which reduces robustness to interharmonics and noise. In addition, the bandwidth increases for smaller values of $|TVG|$.

Figure 12 shows the frequency responses of the transmission functions $T_{3,0}(z)$ and $T_{3,0}^{Q}(z)$ for different values of given parameters (a) l_m^{SB} and (b) $|TVG|_{max}$, in the case of $K = 1$. In this case, the total $|TVG|$ is smaller than in the case of $K = 2$. The order of the compensation filter $N_Q = 16$ is smaller, so the bandwidth is narrower. In the case of $l_3^{SB} = 0.001$, the optimization problem has no solution.

5. Conclusions

CR-based algorithms for harmonic analysis and estimation of harmonic phasors are described in this chapter. The resulting filters for extracting harmonic signals can be of the FIR or IIR type. Algorithms for the optimization of frequency responses are presented and corresponding examples of synthesis are given. Linearized mathematical models were used, which enabled the use of linear optimization methods such as LP and CLLS. When designing the IIR analyzers, a linearized iteration scheme based on the Rouche's theorem was used to control the stability of the system. As for the optimization algorithms, they can potentially be improved by various modifications such as, for example, by nesting optimization loops related to different constraint conditions and/or objectives, and adaptation of iteration steps. It is notable that approximating result could be obtained heuristically thanks to the characteristic position of the pole and zeros. In addition, it seems that closed-form calculation expressions derivation could be possible. On the other hand, the FIR-type algorithm particularly optimizes frequency responses through the postprocessing compensation FIR filters applied to the output signals obtained by the CR structure. This approach allows the usage of a set of compensation filters corresponding to different signal dynamics.

Conflict of interest

The author declares no conflict of interest.

Author details

Miodrag D. Kušljević
Termoelektro Enel ad, Belgrade, Serbia

*Address all correspondence to: miodrag.kusljevic@te-enel.rs

References

[1] Jain SK, Singh SN. Harmonics estimation in emerging power system: Key issues and challenges. Electric Power Systems Research. 2011;**81**:1754-1766. DOI: 10.1016/j.epsr.2011.05.004

[2] Chen CI, Chen YC. Comparative study of harmonic and interharmonic estimation methods for stationary and time-varying signals. IEEE Transactions on Industrial Electronics. 2014;**61**: 397-404. DOI: 10.1109/ TIE.2013.2242419

[3] Phadke AG, Kasztenny B. Synchronized phasor and frequency measurement under transient conditions. IEEE Transactions on Power Delivery. 2009;**24**:89-95

[4] Mai RK, He ZY, Fu L, Kirby B, Bo ZQ. A dynamic synchrophasor estimation algorithm for online application. IEEE Transactions on Power Delivery. 2010; **25**:570-578

[5] de la O Serna JA. Dynamic phasor estimates for power system oscillations. IEEE Transactions on Instrumentation and Measurement. 2007;**56**:1648-1657

[6] Platas-Garza MA, de La O Serna JA. Dynamic phasor and frequency estimates through maximally flat differentiators. IEEE Transactions on Instrumentation and Measurement. 2010;**59**:1803-1811

[7] Barchi G, MacIi D, Petri D. Synchrophasor estimators accuracy: A comparative analysis. IEEE Transactions on Instrumentation and Measurement. 2013;**62**:963-973

[8] Belega D, MacIi D, Petri D. Fast synchrophasor estimation by means of frequency-domain and time-domain algorithms. IEEE Transactions on Instrumentation and Measurement. 2014;**63**:388-401

[9] Platas-Garza MA, de La O Serna JA. Dynamic harmonic analysis through Taylor-Fourier transform. IEEE Transactions on Instrumentation and Measurement. 2011;**60**:804-813

[10] Castello P, Lixia M, Muscas C, Pegoraro PA. Impact of the model on the accuracy of synchrophasor measurement. IEEE Transactions on Instrumentation and Measurement. 2012;**61**:2179-2188. DOI: 10.1109/TIM.2012.2193699

[11] de La O Serna JA. Taylor-fourier analysis of blood pressure oscillometric waveforms. IEEE Transactions on Instrumentation and Measurement. 2013; **62**:2511-2518. DOI: 10.1109/ TIM.2013.2258245

[12] Peceli G, Simon G. Generalization of the frequency sampling method. IEEE Instrumentation and Measurement Technology Conference. 1996:339-343

[13] Kušljević MD, Tomić JJ. Multiple-resonator-based power system Taylor-Fourier harmonic analysis. IEEE Transactions on Instrumentation and Measurement. 2015;**64**:554-563

[14] Péceli G. Resonator-Based Digital Filters. IEEE Transactions on Circuits and Systems. 1989;**36**:156-159

[15] Korać VZ, Kušljević MD. Cascaded-dispersed-resonator-based off-nominal-frequency harmonics filtering. IEEE Transactions on Instrumentation and Measurement. 2021;**70**:Art. no. 1501203. DOI: 10.1109/TIM.2020.3035396

[16] Kušljević MD. Quasi multiple-resonator-based harmonic analysis. Measurement (Lond). 2016;**94**:471-473

[17] Kušljević MD. Adaptive resonator-based method for power system harmonic analysis. IET Science, Measurement and Technology. 2008;**2**: 177-185. DOI: 10.1049/iet-smt:20070068

[18] Chottera AT, Jullien GA. A linear programming approach to recursive digital filter design with linear phase. IEEE Transactions on Circuits and Systems. 1982;**29**:139-149. DOI: 10.1109/TCS.1982.1085123

[19] Tseng CC, Lee SL. Minimax design of stable IIR digital filter with prescribed magnitude and phase responses. IEEE Transactions on Circuits and Systems I: Fundamental Theory and Applications. 2002;**49**:547-551. DOI: 10.1109/81.995676

[20] Messina F, Vega LR, Marchi P, Galarza CG. Optimal differentiator filter banks for PMUs and their feasibility limits. IEEE Transactions on Instrumentation and Measurement. 2017;**66**:2948-2956. DOI: 10.1109/TIM.2017.2728378

[21] Kušljević MD, Vujičić VV. Design of digital constrained linear least-squares multiple-resonator-based harmonic filtering. Acoustics. 2022;**2022**:111-122

[22] Kušljević MD, Tomić JJ, Poljak PD. Constrained-group-delay-optimized multiple-resonator-based harmonic analysis. Tehnicki Vjesnik. 2021;**28**: 1244-1252. DOI: 10.17559/TV-20180327173716

[23] Tomić JJ, Poljak PD, Kušljević MD. Frequency-response-controlled multiple-resonator-based harmonic analysis. Electronics Letters. 2018;**54**: 202-204. DOI: 10.1049/el.2017.4180

[24] Kušljević MD. Determination of minimax design of multiple-resonator-based harmonic estimators through linear programming. Research Trends and Challenges in Physical Science. 2021;**5**:46-62. DOI: 10.9734/bpi/rtcps/v5/5333f

[25] Kušljević MD. Multiple-resonator-based harmonic analysis. Proceedings of 2nd International Conference on Electrical, Electronic and Computing Engineering IcETRAN. Silver Lake: Serbia; 2015. p. MLI1.1.1-12

[26] Kušljević MD, Tomić JJ, Poljak PD. Maximally flat-frequency-response multiple-resonator-based harmonic analysis. IEEE Transactions on Instrumentation and Measurement. 2017;**66**:3387-3398

[27] Kušljević MD. IIR cascaded-resonator-based filter design for recursive frequency analysis. IEEE Transactions on Circuits and Systems II: Express Briefs. 2022;**69**:3939-3943

[28] Kennedy HL. Digital filter designs for recursive frequency analysis. Journal of Circuits, Systems and Computers. 2016;**25**: Art. no. 1630001. DOI: 10.1142/S0218126616300014

[29] Vârkonyi-Koczy AR. Efficient polyphase DFT filter banks with fading memory. IEEE Transactions on Circuits and Systems II: Analog and Digital Signal Processing. 1997;**44**:670-673. DOI: 10.1109/82.618043

[30] Lang MC. Least-squares design of IIR filters with prescribed magnitude and phase responses and a pole radius constraint. IEEE Transactions on Signal Processing. 2000;**48**:3109-3121. DOI: 10.1109/78.875468

[31] Kušljević MD, Tomić JJ, Poljak PD. On multiple-resonator-based implementation of IEC/IEEE standard P-class compliant PMUs. Energies (Basel). 2021;**14**:198. DOI: 10.3390/en14010198